AUDEL®

PLUMBERS AND PIPE FITTERS LIBRARY
Volume I

Materials • Tools • Roughing-In

by Charles McConnell
revised by Tom Philbin

Macmillan Publishing Company
New York
Collier Macmillan Publishers
London

THIRD EDITION
THIRD PRINTING

Macmillan Publishing Company
866 Third Avenue, New York, N.Y. 10022
Collier Macmillan Canada, Inc.

Library of Congress Cataloging-in-Publication Data

McConnell, Charles.
Plumbers and pipe fitters library.

Includes indexes.
Contents: v. 1. Materials, tools, roughing-in—v. 2. Welding, heating, air conditioning—v. 3. Water supply, drainage, calculations.
1. Plumbing—Handbooks, manuals, etc. 2. Pipe-fitting—Handbooks, manuals, etc. I. Philbin, Tom, 1934- . II. Title. III. Title: Plumbers and pipefitters library.
[TH6125.M4 1986] 696′.1 86-8743
ISBN 0-672-23385-1 (v. 1)

10 9 8 7 6 5 4 3

Printed in the United States of America

Foreword

Plumbing and pipe fitting play a major role in the construction of every residential, commercial, and industrial building. Of all the building trades, none are as essential to the health and well-being of the community in general, and the building occupants in particular, as the plumbing trade. It is the obligation and responsibility of every plumber to uphold this vital trust placed in him in the installation of the plumbing materials and equipment.

Every plumbing installation is governed by certain rules and regulations set forth in local plumbing codes that have been adopted from standards established at a local, state, or federal level. In addition, each installation is subject to inspection by a licensed inspector to ensure that all rules and regulations have been complied with. All this, coupled with the usual practice of requiring those persons engaged in plumbing installation to pass an examination for a license, shows the great importance placed on this phase of the building trades.

This series of books has been written to aid persons who wish to become plumbers, as well as those already actively engaged in this occupation and those nonprofessionals who work around their homes. This, the first of three volumes, deals with the tools and materials of the plumbing trade, as well as the various types of pipe joints, pipe fittings, fixtures, and valves and faucets. Also included are helpful chapters describing the correct methods of using blueprints, roughing-in, and pipe fitting. This volume should be helpful particularly to anyone desiring to obtain a license.

Contents

CHAPTER 1

Plumbing Materials 7

Cast iron—steel—copper—brass—lead—tin—oakum—asphaltum—pipe—steel pipe—threaded pipe joints—cast-iron pipe-copper pipe and tubing—brass pipe—tubes—plastic—glass pipe

CHAPTER 2

Basic Tools 31

Rules—measuring tapes—marking or scratch awls—levels—plumb bobs—saws—scrapers—chisels—files—vise stand and chain vise—tube cutters and reamers—pipe cutters—pipe reamer—pipe-bending tools—flaring tools—pipe threaders—pipe and rod threading machines—T-handle tap wrenches—pipe and bolt taps—tap drill sizes—pliers—pipe wrenches—basin wrench—adjustable wrenches—power tools—portable electric drain cleaner—soldering tools

CHAPTER 3

Pipe Fittings 61

Definitions—cast-iron fittings—malleable-iron fittings—brass fittings—cast-steel fittings—forged-steel fittings—various fittings

CHAPTER 4

Pipe Joints 99

Soil pipe—lead and oakum joints—cutting soil pipe—the compression joint—the NO-HUB® joint—coated cast-iron pipe—corrosive waste pipe

CHAPTER 5

BLUEPRINTS 119
Mechanical plans—elevations

CHAPTER 6

PIPE FITTING 137
Pipe cutting—pipe threading—pipe tapping—calculating offsets—assembly—pipe fitting examples—pipe supports—pipe expansion—corrosion—pipe sizes—copper tubing—plastic pipe

CHAPTER 7

ROUGHING-IN 169
Rough-in drawing—isometric drawing—material—heating system rough-in

CHAPTER 8

FIXTURES 213
Lavatories—sinks—bathubs—toilets—urinals—bidet

CHAPTER 9

VALVES AND FAUCETS 259
Valves—faucets—cocks

INDEX 291

CHAPTER 1

Plumbing Materials

Plumbers and pipe fitters come in contact with a wide variety of materials, such as steel, cast iron, copper, brass, and lead, as well as certain chemicals and acids, in the course of their work. Knowledge of these materials and their uses is essential for the proper performance of this work.

Plastics used for pipe, pipe fittings, and fixtures should be added to this list, as should glass used for pipe and pipe fittings. The use of plastics has made a great impact on the piping trades. Advances in technology that led to the introduction of plastics have enabled the piping trades to keep the ever-escalating costs of building within bounds, and at the same time actually to improve the quality of the finished work.

CAST IRON

Cast iron is iron containing so much carbon that it is not malleable at any temperature. It consists of a mixture and combination of iron and carbon, with other substances in varying proportions. Generally, commercial cast iron has between 3 and 4 percent carbon. The carbon may be present as graphite, as in gray cast iron, or in the form of combined carbon, as in white cast iron. In most cases, the carbon is present in both forms. Besides carbon, a combination of silica, sulfur, manganese, and phosphorus are nearly always present.

Cast-iron piping is used for soil, waste, and vent piping and fittings; for water-main piping and fittings; and for natural or manufactured gas mains. Cast iron is used in valves, pipe fittings, and specialty items, which plumbers and pipe fitters use in their work.

STEEL

Steel is an important construction material. Its great strength permits its application to the largest and most severely strained constructive members. It can be forged or cast in any convenient form and is readily obtained in the form of plates, bars, and other shapes. A disadvantage is that it is vulnerable to rust and corrosion, requiring systematic and careful attention in order to preserve it against the action of moisture, oxygen, and carbonic acid.

Upon immersion in a polarizing fluid it is also attacked by galvanic action, in connection with copper or brass. In regard to its percentage of carbon, steel occupies a middle position between cast iron and wrought iron. In common with the former, it has a sufficiently low melting point for casting and, in common with the latter, a sufficient toughness for forging.

According to their varying percentages of carbon, three kinds of steel may be recognized.

1. Soft steel.
2. Medium steel.
3. Hard steel.

Soft steel is nearest to wrought iron in carbon percentage and qualities; it is soft, readily forged, and by careful handling may also be welded. It is principally used in flanged parts, furnace plates, rivets, and other details exposed to alternate heating and cooling or to severe treatment by shaping and forming. Medium steel is harder than soft steel and is used for boiler shells. Cast steel has about the same percentage of carbon as soft or medium steel. In addition, it has silicon and manganese, which are needed to produce good castings. Hard steel comes the nearest to cast iron in carbon percentage and possesses as its most important quality a decided facility for tempering and hardening when cooled quickly in water.

COPPER

Copper is a common metal of a brownish red color, both ductile and malleable and very tenacious. It is one of the best conductors of heat and electricity. It is one of the most useful metals in itself and in its various alloys, such as brass and bronze. It is one of many metals that occur native. It is also found in various ores, of which the most important are chalcopyrite, chalcocite, cuprite, and malachite. Mixed with tin, it forms bell metal; with a smaller proportion, bronze; and with zinc, it forms brass, pinchbeck, and other alloys.

The strength of copper decreases rapidly with a rise of temperature above 400°F; between 800° and 900°, its strength is reduced to about half that at ordinary temperatures. Copper is not easily welded, but may be readily brazed. At near the melting point, it oxidizes (or is burned, as it is called) and loses most of its strength, becoming brittle when cool.

BRASS

Brass is a yellow alloy composed of copper and zinc in various proportions. In some grades, small amounts of tin or lead are added. Brass is used largely for steam and plumbing fittings, electrical devices, builders' hardware, musical instruments, and so on. When zinc is present in small percentages the color of brass in nearly red; ordinary brass for piping contains from 30 to 40 percent zinc. Brass can be readily cast, rolled into sheets, or drawn into tubes, rods, and wire of small diameter. The composition of brass is determined approximately by its color: red contains 5 percent zinc; bronze, 10 percent; light orange, 15 percent; greenish yellow, 20 percent; yellow, 30 percent; yellowish white, 60 percent. The so-called low brasses contain 37 to 45 percent zinc and are suitable for hot rolling; the high brasses contain from 30 to 40 percent zinc and are suitable for cold rolling.

LEAD

Lead was the all-important plumbing material a number of years ago, but it has since been largely replaced by other metals. Lead may be described as a bluish gray metal with a bright luster when melted or newly cut. It is the heaviest of all common metals, weighing .4106 lbs. per cu. in. Commercial lead has a lower specific gravity than pure lead (11.37) because of the impurities contained in it.

The safe working strength of lead is about one-fourth of its elastic limit, or 225 lbs. per sq. in. It is very soft, especially when allowed to cool and solidify slowly. Lead does not crystallize readily. When refined, lead is poured at the correct temperature into a warm mold and allowed to cool. Fernlike crystalline aggregates appear at the surface. In the form of filings, lead becomes a solid mass if subjected to a pressure of 13 tons per sq. in., and liquefies at 2½ times this pressure.

Lead undergoes no change in dry air or in water that is free from air. It becomes pasty at about 617°F and melts at about 650°F. It boils at about 15000°C, but cannot be distilled; its coefficient of linear expansion at ordinary temperatures is .00001571 per 1°F. The strength of lead is both compression and tension is very small. Since lead unites readily with almost all other metals, it is used in many alloys for bearing metals, solders, and the like. Alloys composed of lead, bismuth, and tin are noted for their low melting points.

Lead Poisoning

Lead is a poisonous metal, and the following precautions should be taken in working with this metal to guard against poisoning.

1. Wash your hands carefully before eating or before handling tobacco or anything else that will be placed in the mouth.
2. Bathe frequently.
3. Either change your clothing before going to work or put on outside overalls and jumper while at work. This outside clothing should be washed as frequently as possible.
4. Eat a substantial meal before going to work. With an empty stomach, conditions are more favorable for absorbtion of lead by the body.
5. Drink water and milk plentifully.
6. If you feel at all sick, consult a doctor at once.

TIN

Tin is a soft white metal with a tinge of yellow. It has a high luster, hence is frequently used as reflectors of light. Tin, when nearly pure, has a specific gravity of 7.28 to 7.4, the pure tin being the lightest. It has a low tenacity but is very malleable and can be rolled or laminated into very thin sheets, known as tinfoil. The melting point of tin is 433°F. At 212°F (the boiling point of water) it is ductile and easily drawn into wire. It boils at white heat. It burns with a brilliant white light when raised to a high temperature and exposed to the air. Its specific heat is .0562; latent heat of fusion, 25.65 btu per lb.

Conductivity is low, and it oxidizes slowly in the air at ordinary temperature. When exposed to extreme cold, tin becomes crystalline. Heat conductivity is 14.5; electric conductivity is 12.4 as compared with silver, which is 100. Its weight is 459 lbs. per cu. ft. Tensile strength is 3500 lbs. per sq. in.; crushing load (cast tin) is 15,500 lbs. per sq. in. Due to its high power of resistance to tarnishing by exposure to air and moisture, tin is used as a protective coating for iron and copper. Diluted sulfuric acid has no action on tin when cold, but when tin is boiled in concentrated acid, the metal is dissolved. Coefficient of expansion of tin is .0000151 per 1°F. The principal use of tin by plumbers is for alloying with lead to make solders.

MALLEABLE IRON

The method of producing malleable iron is to convert the combined carbon of white cast iron into an amorphous uncombined condition by heating the white cast iron to a temperature somewhere between 1380° and 2000°F. The iron (sometimes called castings) is packed in retorts or annealing pots, together with an oxide of iron (usually hematite ore). The oxygen in the ore absorbs the carbon in the iron, giving the latter a steel-like nature.

An annealing furnace or oven is used for heating, and the castings are kept red hot for several days or several weeks, depending on the pieces. In order for the process to be successful, the iron must have nearly all the carbon in the combined state and must be low in sulfur. Usually, only good charcoal-melted iron that is low in sulfur is used, although a coke-melted iron is suitable, provided the proportion of sulfur is small. The process is not adapted to very large castings because they cool slowly and usually show a considerable proportion of graphite. Malleable iron pipe fittings are one use of this metal.

OAKUM

Oakum consists of shredded rope or hemp fiber. It is sold a dry oakum for packing poured joints in water mains and as tarred or oiled oakum for use in poured joints of soil pipe. White oakum is a fibrous material covered with a thin woven coating and impregnated with a cementlike powder. White oakum swells when brought into contact with water and is excellent for use as a packing material when making soil-pipe joints.

ASPHALTUM

The name asphaltum is given to a waterproofing paint made from asphalt. Asphalt is black or dark brown in color and will melt or burn, leaving little residue. It dissolves in petroleum or turpentine. It is used for coating pipes and other metals exposed to dampness and weather.

PIPE

Cast iron, steel, brass, copper, and lead pipe have been used by the plumbing and pipe fitting industry for many years. Modern technology has also developed plastics and glass, which are phasing out existing materials.

THREADED PIPE JOINTS

In screwed joints having strength and durability, it is necessary to have clean-cut and uniform threads. The threads should be tapered and smooth, cut with the correct taper, thread angle, and diameters. Reasonable manufacturing tolerances are allowed on all thread elements to take care of variations in threading. The diameters should be such as to allow sufficient hand engagement and yet allow enough threads for wrench or power makeup.

In making up screwed pipe joints, it is very important that the threads in both parts be thoroughly clean. Any threads that may have become burred or bent should be straightened or removed, and a good grade of lubricant should be applied to the threads. The lubricant reduces the friction, which allows the two parts to be pulled up farther, resulting in a more effective pipe joint.

The normal number of external threads to provide a tight joint for various sizes of pipe is given in Table 1-1. These dimensions have been established from tests made under practical working conditions. In order to obtain the correct thread length listed in the table, it is necessary to vary the torque or power according to the size, metal, and weight of the material used. For example, it requires considerably less power to make up a screwed joint using a light bronze valve than a high-pressure steel valve.

Table 1-1. External Pipe-Thread Dimensions

Pipe Size	Min. Reg. for ext. pipe thread
1/8	3/8
1/4	9/16
3/8	5/8
1/2	3/4
3/4	7/8
1	1
1 1/4	1 1/8
1 1/2	1 5/16
2	1 5/8
2 1/2	1 5/8
3	1 11/16
3 1/2	1 11/16
4	1 3/4
5	1 13/16
6	1 15/16
8	2 1/8
10	2 3/8
12	2 9/16
14	2 11/16
16	2 7/8
18	3 3/32
20	3 9/16
24	3 11/16

Leaky Joints

Leaky joints can usually be traced either to faulty threading or an improper lubricant. Frequently the trouble lies in the thread on the pipe, which may have been cut with dull or improperly adjusted threading tools, resulting in wavy, shaved, rough, or chewed threads. Wavy threads are noticeable both to the eye and touch, due to circumferential waves or longitudinal flats of slightly helical form rather than the desired circular form. Shaved threads appear to have been threaded with two dies, one not matching the other, giving a double-thread appearance at the start of the thread. Rough or chewed threads are noticeable rough and torn. Should the threads have any of these defects, it is possible that leaky joints will result.

In order to adapt steel and wrought-iron pipe to different pressures they are regularly made in three grades of thicknesses, known as:

1. Standard.
2. Extra strong (or heavy).
3. Double extra strong (or heavy).

For the three grades, the outside diameters of the listed sizes remain the same, but the thickness is increased by decreasing the inside diameter. For instance, Fig. 1-1 shows sections of the above three grades of pipe of the same listed size.

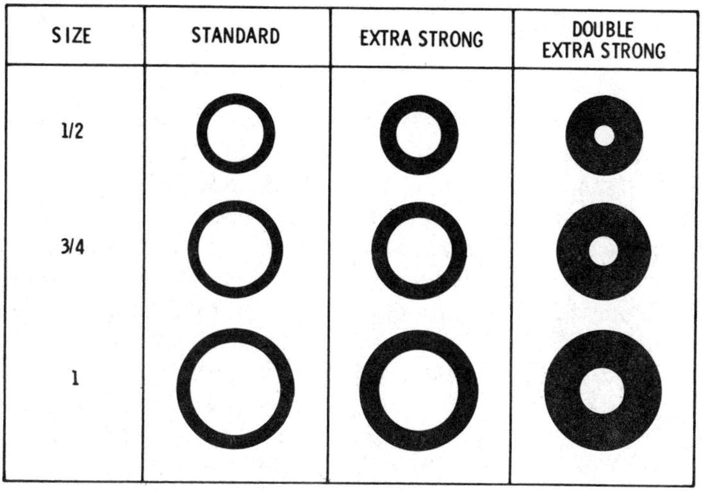

Fig. 1-1. Pipe wall thickness causes inside diameter to vary.

CAST-IRON PIPE

Cast-iron pipe is used extensively by the plumbing and pipe fitting industry. It is used primarily for water mains, gas mains, and soil, waste, and vent piping.

When used for water mains it is connected with poured mineral joints, roll-on and slip-type joints, and bolted mechanical joints, and it can be threaded for use with flanges for flanged joints.

When cast-iron pipe is used for gas mains, bolted mechanical-joint pipe is used.

Cast-iron pipe used for soil, waste, and vent piping is made in several different types: hub and spigot for poured lead joints; NO-HUB® type using stainless steel clamps and neoprene gaskets; and I.P.S. pipe, which can be threaded.

Cast-iron pipe, is extremely long lived, due to its rust-resistant qualities. It can be easily cut to length using hand or hydraulic tools.

COPPER PIPE AND TUBING

Copper pipe and tubing is used extensively in the plumbing and heating/air-conditioning trades. Copper tubing is made in four standard grades of wall thickness: K,L,M, and DWV. The heaviest grade, K, is primarily used for buried or concealed piping. L, a medium weight, is used mostly in accessible areas. M, the lightest weight, is usually used in residential piping and is recommended for use only in accessible areas. K,L, and M tubing is available in lengths (hard). K and L tubing is also made in coils (soft).

DWV, or drainage, waste, and vent, is a medium-weight copper tubing recommended for use in above ground locations. It is a hard-drawn tubing and available in 20 ft. lengths. ACR (air conditioning, refrigeration) tubing is available both in lengths (hard) and coils (soft).

BRASS PIPE

The advantage of brass pipe is that it does not rust or corrode, but it is more expensive than iron pipe. It is made in iron-pipe sizes and is tested to a pressure of 1000 lbs. per sq. in. The temper of the brass is not strictly hard but just sufficiently annealed to prevent cracking and to make it suitable for steam and plumbing work.

Table 1-2. Pipe Test Pressures

STANDARD			EXTRA STRONG			DOUBLE EXTRA STRONG		
Size	Test pressure in pounds		Size	Test pressure in pounds		Size	Test pressure in pounds	
	Butt	Lap		Butt	Lap		Butt	Lap
1/8	700		1/8	700				
1/4	700		1/4	700				
3/8	700		3/8	700				
1/2	700		1/2	700		1/2	700	
3/4	700		3/4	700		3/4	700	
1	700		1	700		1	700	
1 1/4	700	1000	1 1/4	1500		1 1/4	2200	
1 1/2	700	1000	1 1/2	1500	2500	1 1/2	2200	3000
2	700	1000	2	1500	2500	2	2200	3000
2 1/2	800	1000	2 1/2	1500	2000	2 1/2	2200	3000
3	800	1000	3	1500	2000	3		3000
3 1/2		1000	3 1/2		2000	3 1/2		2500
4		1000	4		2000	4		2500
4 1/2		1000	4 1/2		1800	4 1/2		2000
5		1000	5		1800	5		2000
6		1000	6		1800	6		2000
7		1000	7		1500	7		2000
8		800	8		1500	8		2000
8		1000				8		
9		900	9		1500	9		
10		600	10		1200	10		
10		800				10		
10		900				10		
11		800	11		1100	11		
12		600	12		1100	12		
12		800				12		
13		700	13		1000	13		
14		700	14		1000	14		
15		600	15		1000	15		

TUBES

In distinction, a tube has relatively thin walls, and the listed sizes correspond to the outside diameter; a pipe has relatively thick walls, and the listed sizes of wrought pipe do not correspond to the outer diameter. The properties of a 1-inch tube and 1-inch wrought pipe in Table 1-3 will clearly illustrate the distinction between tubes and pipes.

Table 1-3. Properties of 1-Inch Tube and Pipe

	Outside Diameter (Inches)	Inside Diameter (Inches)	Thickness of Metal (Inches)
1-inch tube	1	0.81	0.095
1-inch wrought pipe	1.315	1.049	0.135

PLASTIC PIPING

In recent years, the use of plastic pipe has increased tremendously. Originally it was used primarily for farm water systems and some lawn and golf course underground sprinkling system. Now plastic pipe is in use for natural-gas distribution and supply, chemical and food processing, laboratory and industrial waste disposal, industrial and residential plumbing, and for many other applications. Manufacturers have developed reliable and comprehensive information concerning plastic pipe, recommended methods for joining it, tips on fabrication and installation, and technical data (Fig. 1-2).

Fig. 1-2. Plastic pipe, such as CPVC for hot water shown here, has had a tremendous impact on plumbing trades.

Plastic piping is a unique combination of chemical and physical properties that makes it available at a fairly reasonable cost. In manufacturing plastic piping products, it is essential that only virgin pG lastic compounds that meet exacting specifications be used. In following this practice, the special properties of the raw materials will not be changed or diluted. Products in manufacture must be closely tested and inspected to make certain that each meets or exceeds all applicable standards and specifications. If this practice of extra attention and care at the manufacturing plant is followed, the plumber and pipe fitter will know that their field work will be made easier and the completed job will be satisfactory.

The term *plastic* includes a number of materials that differ significantly in their properties, characteristics, and suitability for specific jobs. These differences are important in avoiding misapplications. Plastic materials generally are classified in two basic groups—thermoplastics and thermosetting resins. The thermoplastics can be reformed repeatedly by application of heat. Thermosetting resins, once their shape is fixed and cured, cannot be changed for reuse.

Thermoplastics material is used in the greater portion of the plastic pipe manufactured. A variety of end treatments allows the fabricator a choice of joining pipe to fittings by solvent welding, threading, fusion welding, flanging, or almost any combination of these methods to assure a satisfactory job, but some joining methods are more suitable than others in certain cases.

Polyvinyl Chloride (PVC)

PVC has a high tensile strength and a good modulus of elasticity. Therefore it is stronger and more rigid than most other thermoplastics. The maximum service temperature is 150°F for Type I (Normal Impact), and 140°F for Type II (High Impact). PVC has excellent chemical resistance to a wide range of corrosive fluids, but may be damaged by ketones, aromatics, and some chlorinated hydrocarbons. It has proved to be a good material for processing piping (liquids, gases, and slurries), water service, and industrial and laboratory chemical waste drainage. Drain, waste, and vent piping can be joined by solvent welding or fillet welding. A close relative is CPVC, which is used as hot-water pipe.

Polyvinyl Dichloride (PVDC)

PVDC is particularly useful for handling corrosive fluids at temperatures 40° to 60° above the limits for other vinyl plastics. Suggested uses include process piping for hot corrosive liquids, hot- and cold-water lines in office buildings and residences, and similar applications above the temperature range of PVC. PVDC pipe should be joined by solvent welding, threading, or as recommended by the pipe manufacturer.

Polypropylene

Polypropylene is the lightest of the thermoplastic piping materials, yet has higher strength and better general chemical resistance than polyethylene and may be used at temperatures 30° to 50° above the recommended limits of polyethylene. Polypropylene is an excellent material for laboratory and industrial drainage pipe where mixtures of acids and solvents are involved. It has found wide application in the petroleum industry where its resistance to sulfur-bearing compounds is particularly useful in salt-water disposal lines, low-pressure gas-gathering systems, and crude-oil flow piping. It is best joined by fusion welding.

Polyethylene

Polyethylene is the least expensive of the thermoplastics and one of the most widely used. Although its mechanical strength is comparatively low, it exhibits very good chemical resistance and is generally satisfactory when used at temperatures below 120°F. Types I and II (low- and medium-density) are used frequently in chemical laboratory drainage lines, field irrigation, and portable water systems. Fusion welding is the best method for joining this material.

Acrylonitrile-Butadiene-Styrene (ABS)

ABS has high impact strength, is very tough, and may be used at temperatures up to 180°F. It has a lower chemical resistance and lower design strength than PVC. ABS is used for carrying water for irrigation, gas transmission, drain lines, waste, and vent piping. Solvent welding and threading are recommended fabrication methods.

Crosslinked Polyethylene (CAB-XL)

CAB-XL is a material that has excellent strength characteristics and improved resistance to most chemicals and solvents at elevated temperatures to 203°F. Crosslinking permits high impact strength even at subzero temperatures. This material is often suggested for services too severe for ordinary polyethylene.

Glass-Reinforced Epoxy

Glass-reinforced epoxy is probably the best thermoset plastic for piping applications. It has a high strength-to-weight ratio and an outstanding resistance to chemicals and weathering.

Plastic-Pipe Connections

A broad range of domestic and industrial thermoplastic and thermoset piping materials are manufactured. Fig. 1-3 indicates

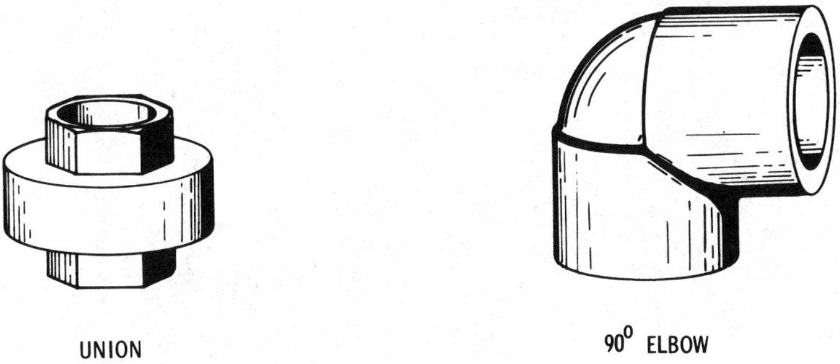

Fig. 1-3. Plastic pressure fittings.

some pressure fittings and pressure valves. Fig. 1-4 illustrates the threaded fitting. Fig. 1-5 shows the reinforced epoxy fittings; Fig. 1-6 shows Cabot Thermo-seal drainage fittings of polyethylene and polypropylene; and Fig. 1-7 illustrates PVC (Polyvinyl Chloride) industrial drainage pipe and fittings.

Handling and Storing

Normal precautions should be used when unloading and storing plastic pipe. Deep scratches and gouges on the pipe surface can lead to reduced pressure-carrying capacity. Standard pipe wrenches can deform or scar threaded plastic pipe when being fabricated. Strap wrenches are recommended. Pipe placed in a pipe vise or chuck should be wrapped at the jaw location with emery cloth or soft metal.

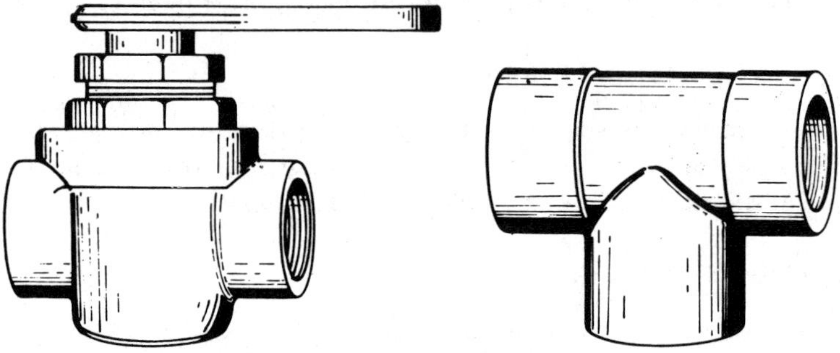

VALVE TEE

Fig. 1-4. Plastic threaded fittings.

90^{0} ELBOW ADAPTOR

Fig. 1-5. Plastic epoxy fittings.

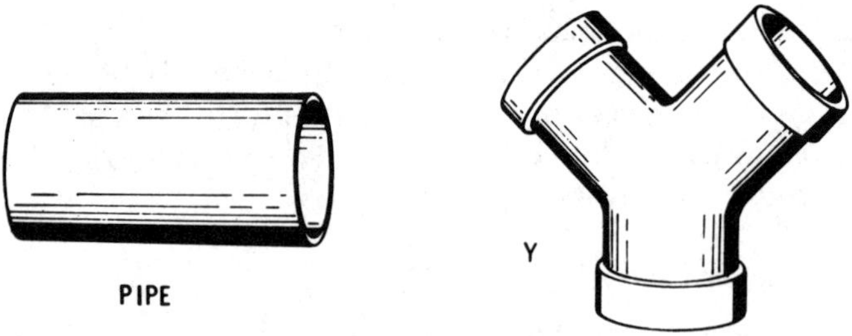

PIPE Y

Fig. 1-6. Thermoseal fittings.

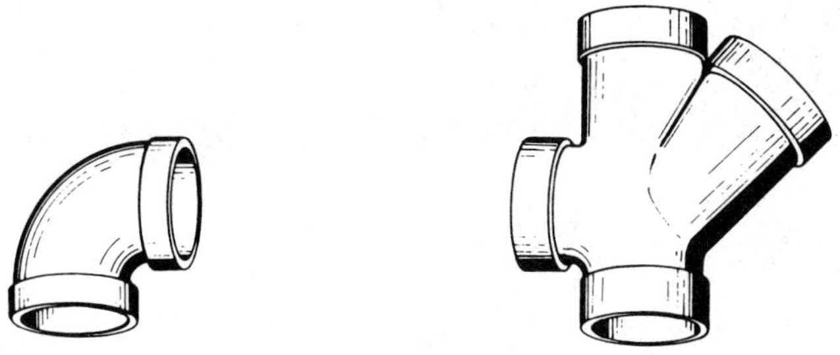

Fig. 1-7. Polyvinyl chloride fittings.

Store pipe in a clean area with adequate ventilation. Do not mix plastic fittings and flanges with metal-pipe components—store separately. Avoid burrs and sharp edges on metal racks. Store pipe on racks that afford continuous support to prevent sagging or draping of long lengths. Do not store or install plastic pipe near steam lines or other heat sources that could overheat or damage the pipe.

Cutting

Tubing cutters equipped with a special cutter wheel for plastics are best to use when cutting plastic pipe. A hacksaw or a handsaw can also be used; if a handsaw is used, it should have at least 12 teeth per inch. A miter box will aid in making a good square cut when a handsaw is used. A half-round smooth file or a deburring tool should be used to remove the burrs or rough edges from the cut. A tubing cutter equipped with a special wheel for cutting plastic tubing is shown in Fig 1-8.

Bending

Bending leaves residual stresses in plastic pipe. The use of bends is not recommended, particularly when the line is to operate at or near maximum rated temperatures and pressures. Factory-made fittings and straight lengths of pipe give better performance.

When field bending is required for special purposes, or to provide for expansion and contraction, the following practices are recommended:

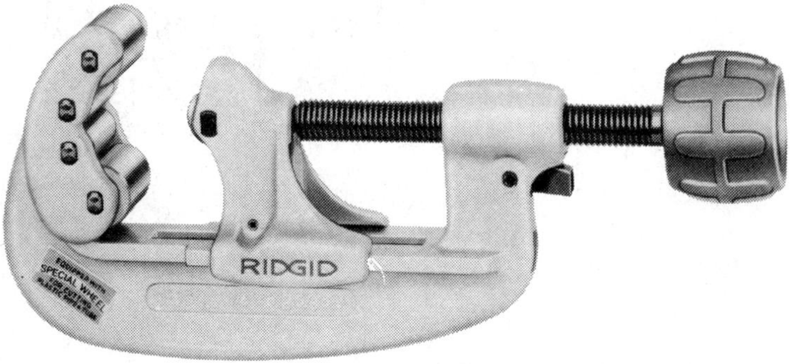

Courtesy Ridge Tool Co.

Fig. 1-8. A tubing cutter with cutter wheel for plastics.

1. Seal both ends of the pipe length with a plumber's test plug and introduce sufficient air pressure to maintain ovality of the pipe during bending. The same purpose can be achieved by filling the pipe with preheated sand.
2. Heat the pipe uniformly by immersing in hot oil or water, or by rotating in front of a hot-air gun. Do not use an open flame.
3. When the pipe becomes soft and pliable, place in a wooden forming jig and bend it as quickly as possible to prevent weakening or deforming. Minimum radius to which a bend should be made is 5 to 6 pipe diameters, but the initial forming bend should be slightly greater to allow for spring-back.
4. Keep bend in forming jig until the pipe cools and becomes rigid and then cool it quickly by immersion in water. Do not remove sand or relieve air pressure until after final cooling.

Thermoset plastics, such as glass-reinforced epoxy, cannot be field bent by heating.

Solvent Welding

A good method for joining rigid thermoplastics, such as PVC and PVDC, is by solvent welding. This method provides stronger and tighter joints than threading. Engineers have suggested the following useful pointers:

1. Use the proper solvent cement—PVC cement with PVC pipe and PVDC cement with CPVC pipe.
2. When solvent welding lighter PVC pipe, apply lightweight PVC cement to the pipe O.D. only, not to the pipe and fitting as is done when solvent welding heavier pipe.
3. Leave a fillet bead between pipe and fitting when solvent welding PVC and CPVC piping.
4. Use a natural (hog) bristle brush for applying solvent cement. Nylon and other synthetic materials are attacked by solvents in the cement.
5. Use a ½-in.-wide brush for pipe ½ in. through 1 in. a 1-in.brush for 1¼-in. through 2-in pipe, and a 2-in. brush for pipe 3-in. and larger.
6. An ordinary oil can is an excellent container for acetone-type cleaner. Excessive evaporation is prevented and the solvent is always handy.
7. Do not allow water to come into contact with solvent cement. Wrap a cloth or handkerchief around the forehead in hot weather to keep perspiration from dripping into the cement. When not using, keep the cement covered.
8. Allow solvent cement to cure 5 to 15 minutes before handling and wait 24 hours before introducing full line pressure into a solvent cemented piping system.
9. At the end of the day, place the brush in cleaner and cover the cement tightly. When reusing a brush, shake excess cleaner out before dipping into the cement.

Fillet Welding

Fillet welding is the generally accepted practice of repairing leaks in thermoplastic piping systems. Plumbers and pipe fitters learn fillet welding rapidly, since general procedures are similar to those used for welding and brazing metals. However, the following general practices are recommended by engineers and technicians.

1. Welding surfaces must be clean of dirt, oil, moisture, and loose particles of plastic material. When welding solvent-welded joints, allow the cement to cure six hours before welding. Remove all excess cement residue with a knife,

emery cloth, or wire brush before welding. Do not weld a joint while it is leaking as the moisture will prevent a good bond.

2. Maintain uniform heat and pressure on the rod while welding. Too much heat will char, melt, or distort the material; too much pressure on the rod tends to stretch the weld bead, which may result in cracks and checks in the weld after it cools.
3. Do not splice welds by overlapping side by side. When terminating a weld, lap the bead on top of itself (not alongside) for a distance of 3/8 to 1/2 in.
4. A single-drip leak usually can be repaired with a single bead weld; more serious leaks require full fillet welds, usually three beads and up to five beads in large-diameter pipe. When making multiple-layer welds, allow sufficient time for each pass to cool before proceeding with final welds.
5. When welding PVC or PVDC, hold the rod at an angle of 90° to the work; when welding polyethylene, polypropylene, and penton, hold the rod about 75° away from the gun, as indicated in Fig. 1-9.

Fig. 1-9. A fillet welding thermoplastic pipe.

Table 1-4 shows a fillet welding chart including the recommended welding temperatures.

Table 1-4. Fillet Thermoplastic Pipe Welding Chart

	PVC Type I	PVC Type II	PVDC	PP	PE	PENTON
Welding Temperature	500°F to 550°F	475°F to 525°F	500°F to 550°F	550°F to 600°F	500°F to 550°F	600°F to 650°F
Welding Gas	Air	Air	Air	Inert	Inert	Air
Odor Under Flame	HCL	HCL	HCL	Wax	Wax	Sweet chlorine
Position of Rod	90°	90°	90°	75°	75°	75°
Remarks			Low weld strengths	May splash; reduce airflow		May splash; reduce airflow

Underground Installation

Depth of the ditch is not necessarily critical except in freezing areas where it is recommended that pipe be laid well below the frost line. If the soil at the trench is unyielding (clay, as an illustration), overexcavate the trench about 4 inches and place a cushion of rock-free soil or coarse sand to bring the trench to final grade. When ground water is encountered, the bottom must be stabilized by overexcavating about 12 inches and filling to the grade with gravel having a maximum particle size of ½ inch. Keep the trench free of ground water during laying and joining, and until it is backfilled sufficiently to prevent flotation of the pipe.

Final trimming and grading must be accomplished by hand. Make sure the trench bottom is smooth and regular to avoid local bending. Do not block the pipe to grade. The 8 to 12 inches of backfill material must be free of rocks and other cutting or sharp objects. Fine sand, clay, silt, and frozen soil lumps are not recommended for backfilling. Assemble plastic pipe in sections above ground and then lower into the trench. It is recommended that several lengths of pipe be placed in the trench bottom so that the pipe will not be cocked or canted while a joint is being made.

When solvent-welded joints are used, it is recommended that backfilling be delayed about five minutes after completion of the joint to allow the cement to take initial set. Pressure testing is often completed prior to backfilling.

During backfilling, it is recommended that the pipe be at operating temperature and pressure if possible. This practice prevents pipe deformation due to expansion. Compensation for thermal expansion in buried plastic pipe can be achieved by snaking the line from side to side in the trench. One cycle for each 40 feet or less is satisfactory in most cases.

Expansion in Plastic Pipe

When total temperature change is less than 30°F, special provisions for accommodating thermal expansion are not generally required, particularly when the line includes several directional changes, which provide inherent flexibility. Exercise caution with threaded connections, however, as they are more vulnerable to failure by bending stresses than are solvent-welded joints. When expansion cannot be assumed by regular dimensional changes, several methods to compensate for expansion are available. One is to fabricate an offset or expansion loop using elbows and straight pipe joined by solvent welding.

Another is the use of an expansion joint. The latter is particularly recommended for large-diameter pipe and where space for offset lines is limited. An expansion joint is basically two tubes, one telescoping inside the other. The other tube is to be firmly anchored and the inner tube allowed to move with a piston-like action as the attached pipe expands or contracts. In long runs (15 feet or more) it is recommended the pipeline be anchored at each change of direction so the expansion movement in the pipe can be directed squarely into the expansion joint.

Alignment of expansion joints is important, as binding may result if the pipe is canted or cocked and does not move in the same plane as the joint. It is recommended that guide loops be installed approximately one floor from each end of each expansion joint. If there is any doubt about expansion joints for plastic pipe, contact a representative of the piping company or the manufacturing company for information.

Supporting Plastic Pipe

Most companies recommended that thermoplastic piping be supported at intervals roughly ½ to ¼ of that normally required for steel pipe. Supports and hangers can be clamp, saddle, angle, spring, or other standard types. Broad and smooth-bearing surfaces are better than narrow or sharp contacts as they minimize danger of stress concentration and physical damage. Continuous support in channel iron often is provided for lines operating at high temperatures or handling hazardous liquids at high temperatures. Angle irons suspended with clevis hangers have also been used successfully. Avoid clamping the pipe so to prevent endwise movement needed to take care of thermal expansion. Rigid clamping is advisable at valves and fittings located near sharp changes in direction when the line is subjected to wide temperature changes.

Do not lay plastic pipe on steam lines or other high-temperature surfaces. With the exception of couplings, support all plastic fittings individually, and brace valves against operating torque. Generally, vertical runs are supported by spring hangers and guided with rings or long U-bolts, which restrict movement of the rise to one plane. It is sometimes helpful to support a long riser with a saddle at the bottom. General residential plastic piping is available for complete piping systems. A plastic drain-waste-vent pipe and fittings system offers many advantages. It is corrosion proof; it cannot rust, corrode, or rot. Ordinary household chemicals and effluents do not affect it. A smooth interior has a tendency to eliminate buildup of deposits. Plastic pipe is also available for installation with septic systems. Before using plastic, or any pipe for that matter, check local plumbing codes.

GLASS PIPE

Glass pipe is widely used in a great variety of industries. Dairy products, food, chemical, pharmaceutical processors, and paper and pulp, find that the corrosion resistance and transparency of glass pipe make it suitable for applications where observation of the processing is vital. Glass piping in short lengths can be inserted into nontransparent piping for use as sight-flow indicators. Glass pipe is also lightweight and the extra smooth inside surface reduces pressure drop due to friction and discourages scale buildup. Matching glass fittings are made for use with the glass pipe, and several different types of flanged connections and gaskets can be selected to fit the particular job need. Glass pipe can be easily cut to length and can be used with plain ends or beaded on the job if the application makes beading desirable.

Another form of glass piping used widely, especially for acid wastes or acid conducting pipe, is a fiberglass pipe. This pipe is made of fiberglass, spun and coated with a resin; it is tough, lightweight, and impervious to most acids.

CHAPTER 2

Basic Tools

The plumber and pipe fitter are expected to know how to use a greater variety of tools than any other building tradesmen. Many specialized tools are needed. The tools listed and illustrated in this chapter are common. Some specialized tools will be pictured and described in chapters describing the work in which they are used.

RULES

A good six-foot rule is a necessary tool for the plumber and pipe fitter. A wooden folding *inside reading* rule (Fig. 2-1) is the most practical rule for several reasons. It can be opened partway, say to the 30-in. mark, and when laid on a set of plans, the rule will lay flat, thus showing more accurate measurements. The numbers from 1 to 6 are used more often than 66 to 72, and the low numbers, being on the inside, are protected from wear.

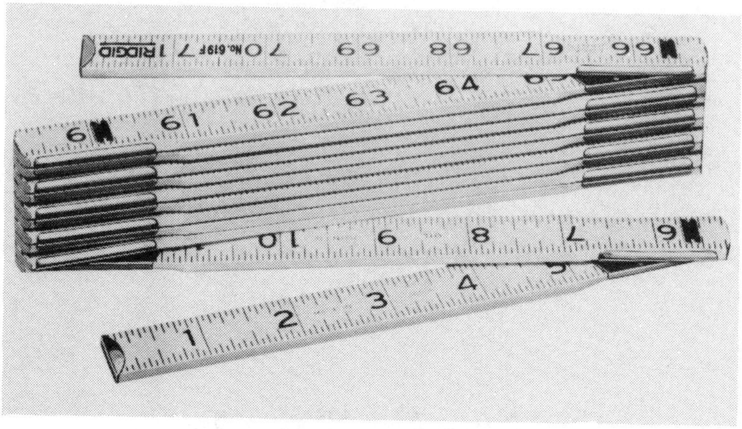

Courtesy Ridge Tool Co.

Fig. 2-1. A wooden folding rule.

MEASURING TAPES

Measuring tapes can be purchased in a variety of lengths and types, but the most common are the 50- and 100-ft. lengths. For measuring distances of 6 to 10 ft., a pocket type with a pushbutton spring wind is available. The longer tapes, as shown in Fig. 2-2, are normally used for measuring pipe for installations requiring long runs.

Fig. 2-2. A 50-ft. steel measuring tape.

Courtesy L.S. Starrett Co.

MARKING OR SCRATCH AWLS

An awl consists of a short piece of round steel, pointed at one end, with the other end fixed in a convenient handle. A scratch awl is an inexpensive *form of scriber* used in laying out fine work where a lead pencil mark would be too coarse for the required degree of precision. The scratch awl is shown in Fig. 2-3.

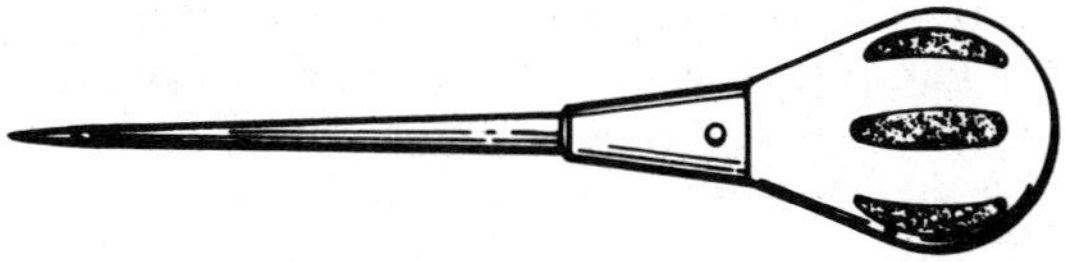

Fig. 2-3. A scratch awl.

LEVELS

A level is used for both guiding and testing: to guide in bringing the work to a horizontal or vertical position, and to test the accuracy and levelness of the completed construction and piping. Levels are made in a variety of types and shapes, but all are used for the same purpose. Fig. 2-4 shows a typical level.

Courtesy Ridge Tool Co.

Fig. 2-4. A level.

PLUMB BOBS

The word *plumb* means *perpendicular to the plane of the horizon*, and since the plane of the horizon is perpendicular to the direction of gravity at any given point, the force due to gravity is utilized to obtain a vertical line in the device known as a plumb bob, shown in Fig. 2-5. The plumb bob is made from solid steel, bored and filled with mercury to provide a low center of gravity and great weight in proportion to its short length and small diameter. A no-roll hex head prevents rolling when the plumb bob is set down. Since the point is removable, it can be easily replaced if broken or worn. An outstanding feature that results in the bob hanging perfectly true is the device for fastening the string, without a knot to tie or untie, by simply drawing it into the specially shaped slotted neck at the top.

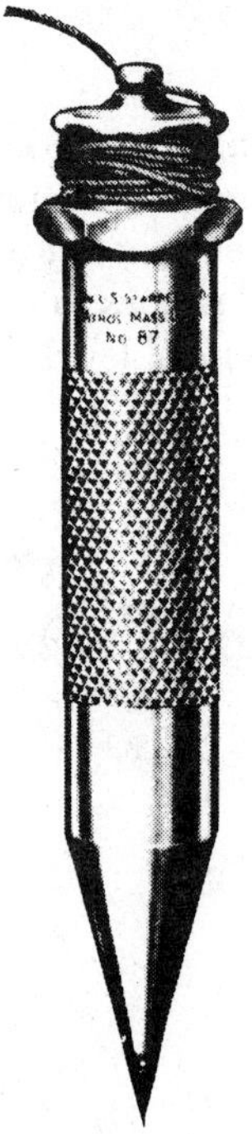

Fig. 2-5. A plumb bob.

Courtesy L.S. Starrett Co.

SAWS

Handsaws such as the one shown in Fig. 2-6 are used by plumbers and pipe fitters primarily for cutting framing lumber when roughing-in a job or for cutting ABS or PVC plastic piping. A saw having 12 teeth per inch is a good all-around saw for this work.

Hacksaws are used by plumbers and pipe fitters for sawing many different materials. There are many occasions when piping must be cut in place and no other tool will serve the purpose. The hacksaw blade in this type saw can be mounted vertically or horizontally for cuts in difficult places. The saw shown in Fig. 2-7 has storage space for six extra blades in the backbone of the frame.

Fig. 2-6. A handsaw.

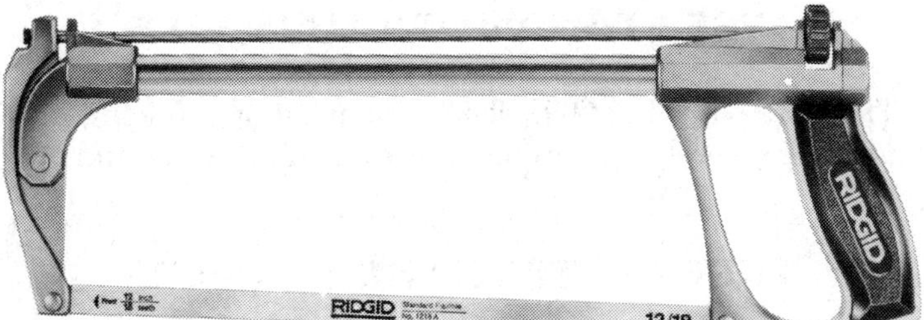

Courtesy Ridge Tool Co.

Fig. 2-7. A hacksaw.

CHISELS

The plumber frequently must use the chisel in cutting away wood members to make room for pipes. A chisel should be absolutely flat on the back (the side not beveled). An inferior chisel is ground off on the back near the cutting edge, with the result that it tends to follow the grain of the wood, splitting it off unevenly because the user cannot properly control the chisel. The flat back allows the chisel to take off the very finest shaving, and where a thick cut is desired, it will not strike too deep. This is a point to be found in good chisels.

Chisels are made of selected steel with the blade almost imperceptibly widening toward the cutting edge. The blades are oil tempered and carefully tested. The ferrule and blade of the socket chisel are so carefully welded together that they form a single piece. Handles are generally of very durable plastic materials, although some chisels have highly finished hickory handles.

FILES

Files (Fig. 2-8) are often used by plumbers and pipe fitters and for a great variety of purposes such as cleaning flange faces, cleaning welds, and removing the burrs from pipe. Smooth files are best for use on copper and brass; coarse files are favored for use on iron and steel.

VISE STAND AND CHAIN VISE

The chain vise (Fig. 2-9), shown mounted on a folding stand, holds pipe securely for cutting, threading, fabrication, and so on.

TUBE CUTTERS AND REAMERS

The tubing cutter shown in Fig. 2-10 is made of lightweight aluminum alloy and can be used to cut copper, aluminum tubing, brass pipe, and thin wall conduit. The cutter shown has a capacity

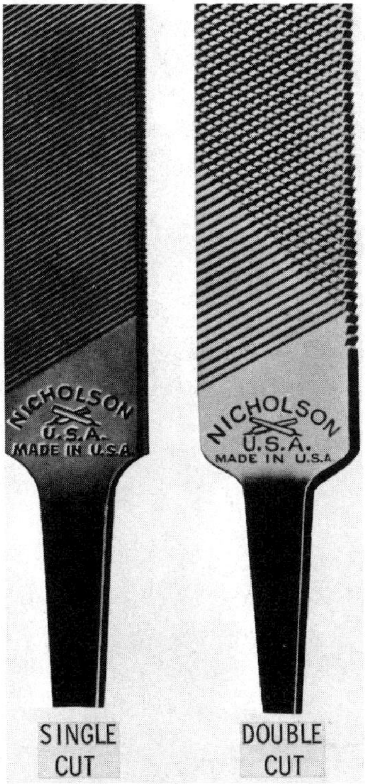

Fig. 2-8. Single- and double-cut files.

of 1/8- through 1 1/8-in. tubing. Similar cutters are used to cut tubing up to and including 4 1/8 in. O.D. Special cutter wheels are available for these cutters for use in cutting plastic tubing. Fig. 2-11 shows a similar tubing cutter used to cut 2- to 4-in. tubing.

PIPE CUTTERS

Fig. 2-12 shows a typical pipe cutter that can be used by hand or with power equipment. It can be converted to a three-wheel cutter for use in tight places by removing the two rollers and replacing them with cutter wheels. It is adjustable to cut pipe from 1/8 in. through 2 in.

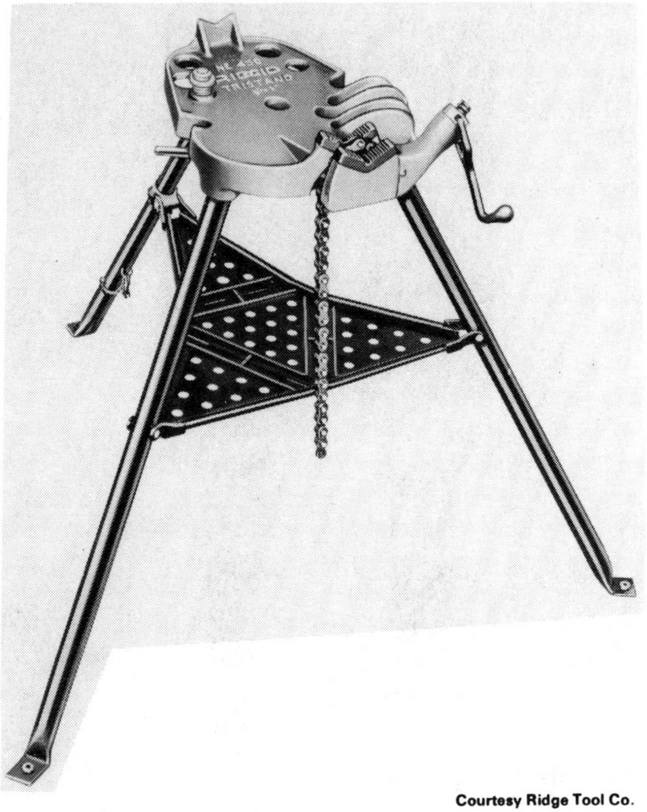

Courtesy Ridge Tool Co.

Fig. 2-9. Vise stand and chain vise.

Fig. 2-10. A tubing cutter with fold-in reamer.

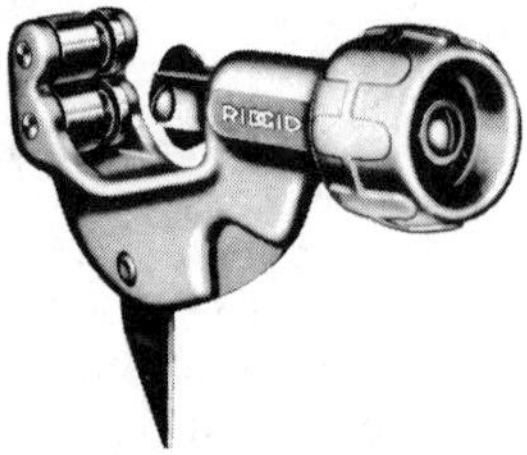

Courtesy Ridge Tool Co.

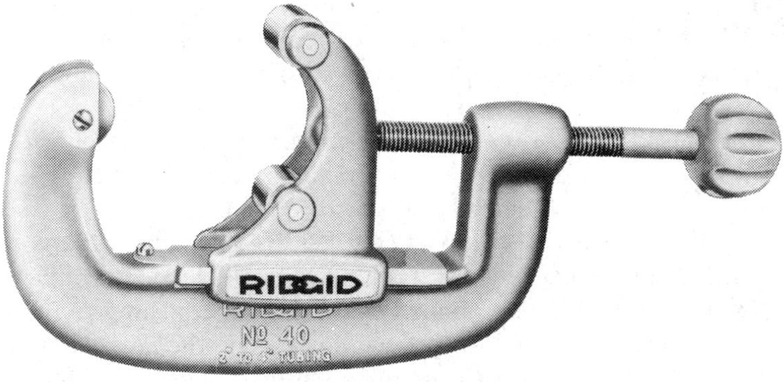

Courtesy Ridge Tool Co.

Fig. 2-11. A tubing cutter used to cut 2- and 4-inch tubing.

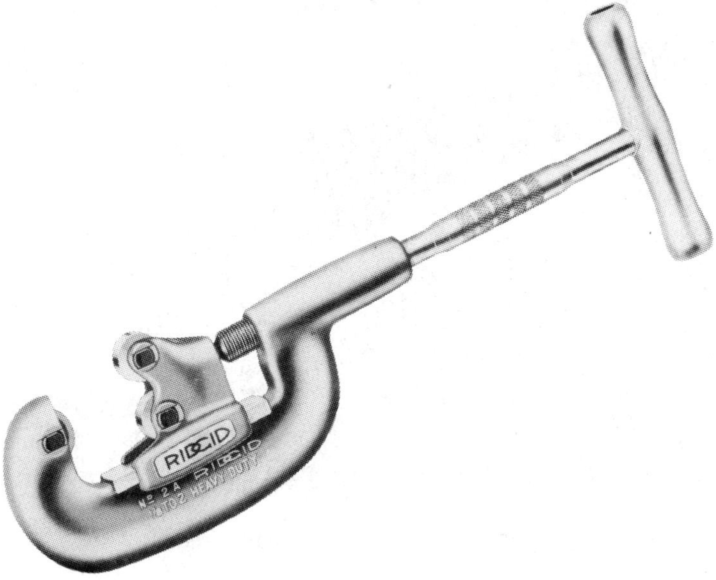

Courtesy Ridge Tool Co.

Fig. 2-12. A pipe cutter.

Soil Pipe Cutters

Tools have been developed that make a hammer and chisel method of cutting soil pipe obsolete. Also, the trend in the industry is to change from extra-heavy soil pipe to the new, light-service-weight soil pipe. For cutting extra-heavy soil pipe, clay tile, cement pipe, and Class 22 water main, a chain cutter such as is shown in Fig. 2-13A is used; for 2- through 6-in. NO-HUB® soil pipe, a cutter such as is shown in Fig. 2-13B is best, due to the closer spacing of the cutter wheels. The operation of both types of cutter is the same; the jaws are opened by turning the adjusting knob, the chain is locked around the pipe and placed in the jaw notch, the ratchet knob is set with the arrow pointing down, and a few easy pumps tighten the chain until the pipe is severed.

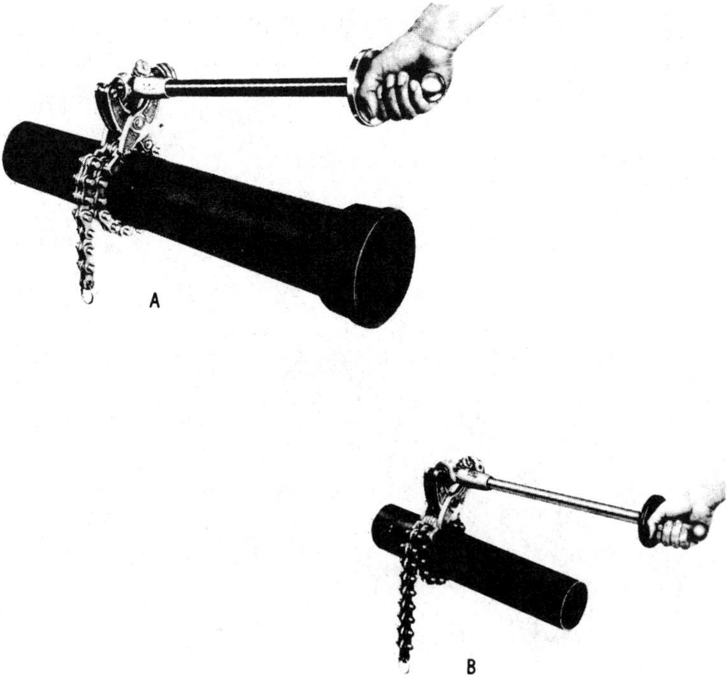

Courtesy Ridge Tool Co.

Fig. 2-13. A chain pipe cutter.

Hinged Pipe Cutter

It is often necessary for plumbers and pipe fitters to cut a section of pipe in a ditch or other tight quarters. The cutter shown in Fig. 2-14 is used primarily for this purpose.

PIPE REAMER

Pipe cut by any method should be reamed to remove inside burrs. The spiral reamer in Fig. 2-15 is self-feeding and can be used by hand or with power equipment.

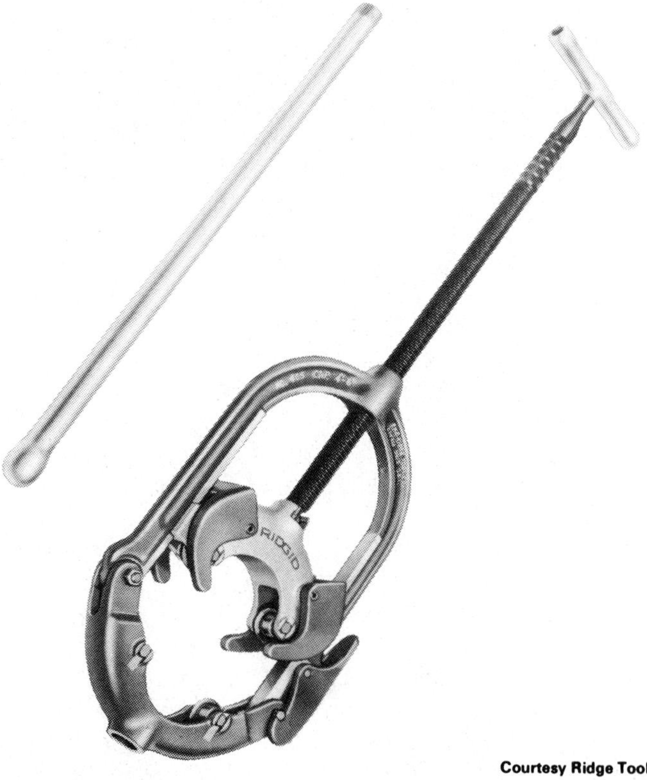

Courtesy Ridge Tool Co.

Fig. 2-14. A hinged pipe cutter.

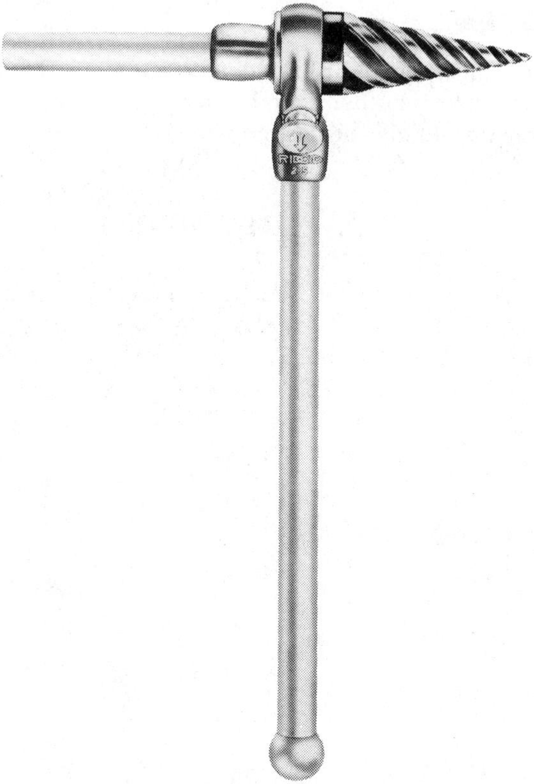

Courtesy Ridge Tool Co.

Fig. 2-15. A pipe reamer.

PIPE-BENDING TOOLS

There are a variety of devices for bending pipe, both power- and hand-operated. One of the most common pipe benders is illustrated in Fig. 2-16. This lever-type bender is versatile, accurate, and easy to use. Correctly measured, bends will be accurate to blueprint dimensions within plus-minus $\frac{1}{32}$ in. A scale on the link eliminates extra measuring and assures fast tube positioning for accurate finished dimensions. To eliminate possible hand injury, handles are held wide apart when making 180° bends. There are six sizes for soft or hard copper, brass, aluminum, steel, and stainless-steel tube.

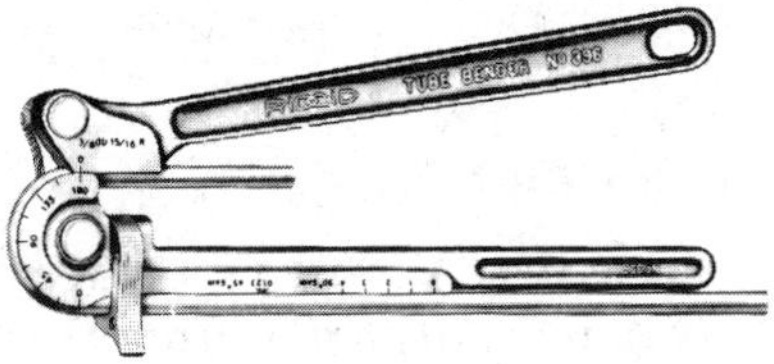

Fig. 2-16. Lever-type tube bender.

FLARING TOOLS

Several different flaring tools are often needed by the plumber and pipe fitter. The type shown in Fig. 2-17A is called a flaring block and will flare several different sizes of tubing. It is made as a unit; the yoke cannot come off and be lost. The feed releases automatically when the flare is completed.

The flaring tool shown in Fig. 2-17B is a hammer-type tool and is usually used for flaring copper water tubing.

A

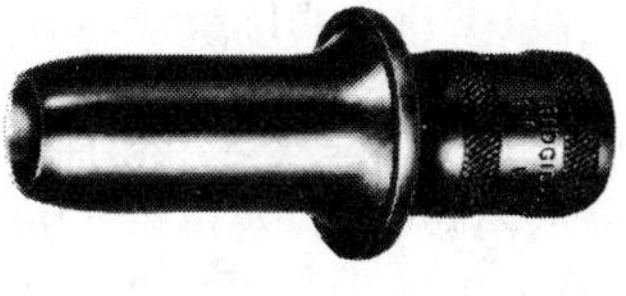

B

Courtesy Ridge Tool Co.

Fig. 2-17. Typical flaring tools.

PIPE THREADERS

⅛ to 1 Inch

The ratchet threader shown in Fig. 2-18 is available with separate heads to thread pipe from ⅛ in. through 1 in. Ratchet-type dies can be used to thread pipe in place, if necessary, and can be used by hand or with power equipment. Similar threaders can be used to thread pipe up to and including 2 in. in size. The heads are removed from the stock by pulling out on the ratchet knob.

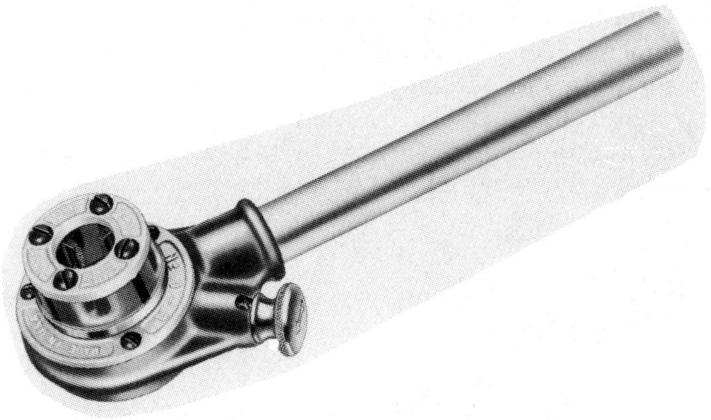

Courtesy Ridge Tool Co.

Fig. 2-18. ⅛- to 1-in. pipe threader.

1 to 2 Inches

The adjustable ratchet threader shown in Fig. 2-19 uses one set of dies to thread pipe from 1 in. up to and including 2 in. Using this type threader, the depth of thread is also adjustable, by varying the starting point of the threading head. There are two steps to threading pipe with this threader. First, the dies and the rear chuck must be set for the pipe size. The rear chuck face should be turned to the pipe-size setting and the locking cam left in open position. To set the dies for the pipe size, turn the rear chuck counterclockwise until the pins (A), which set the die head, are free from the plate (B). When free, the pins can be moved to line up with the proper pipe-size marking. The rear chuck should then be turned clockwise, and the pins will enter the alignment holes in the plate.

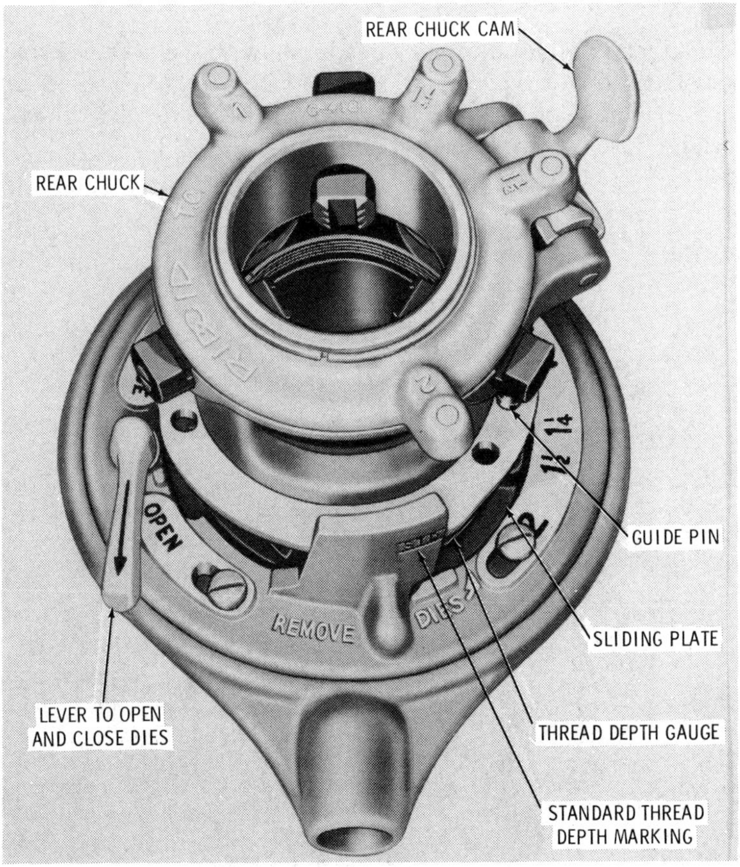

Courtesy Ridge Tool Co.

Fig. 2-19. 1- to 2-in. pipe threader.

Turn the rear chuck until the plate is in line with the *STD* marking on the threader body. Slide the threader onto the pipe, center the pipe against the pipe dies, and lock the cam on the rear chuck. The dies are now ready to cut a standard depth thread. The starting points for deeper or shallower than normal threads are shown in Fig. 2-19.

2 1/2 to 4 Inches

The geared adjustable pipe threader shown in Fig. 2-20 uses one set of dies to thread 2 1/2-, 3-, 3 1/2- and 4-in. pipe. It can be used for hand threading or with a power machine, and the depth of thread can be varied to suit a given condition.

Courtesy Ridge Tool Co.

Fig. 2-20. 2 1/2- to 4-in. pipe threader.

PIPE AND ROD THREADING MACHINES

Pipe machines are used to save both time and labor. The pipe machine shown in Fig. 2-21 will thread pipe from 1/8 in. through 2 in. and rod from 1/4 in. through 2 in. The die heads used are quick opening type; the dies, cutter, reamer and oiler swing out of the way when not in use. The machine can be mounted on a wheeled stand for easy moving.

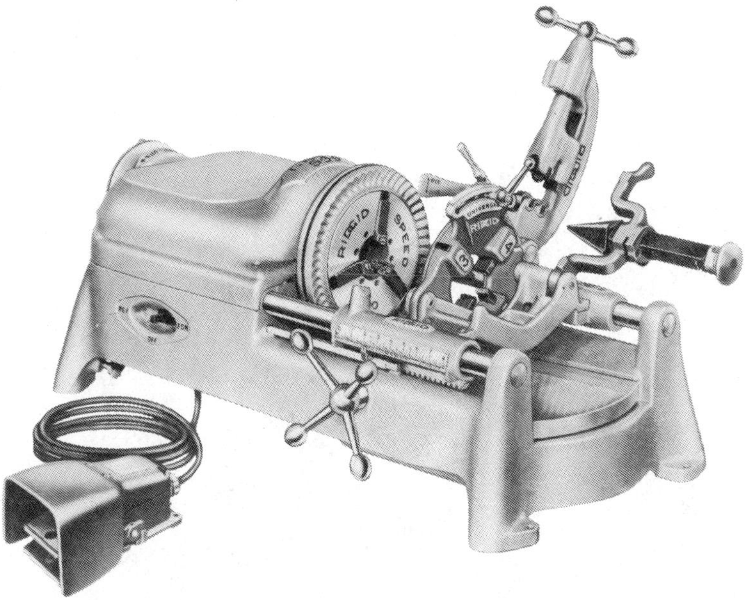

Courtesy Ridge Tool Co.

Fig. 2-21. 1/8- to 2-in. pipe and rod threading machine.

2 1/2- through 4-in. pipe can also be cut and threaded on a pipe machine. The machine shown in Fig. 2-22 uses quick-opening die heads; the reamer, cutter, and oiler swing out of the way when not in use. This machine can either be bench mounted or mounted on a wheeled stand for easy moving.

T-HANDLE TAP WRENCHES

Occasionally it may be necessary to work in close quarters where only a small space is available. T-handle wrenches can be used for this purpose. They are used for holding taps, drills, reamers, and other small tools to be turned by hand. The length of

Courtesy Ridge Tool Co.

Fig. 2-22. 2½- to 4-in. pipe and rod threading machine.

the body and the tape and shank size vary in capacity. Fig. 2-23 shows a tool of this type.

The construction is such that the jaws conform to the tool being held, making it rigid and less apt to become loose. The wrenches have a sliding handle, which is frictionally held. This feature permits the handle to be positioned so that leverage can be applied when working in close quarters. Ratchet-type T-handles are also available, making this wrench even more useful.

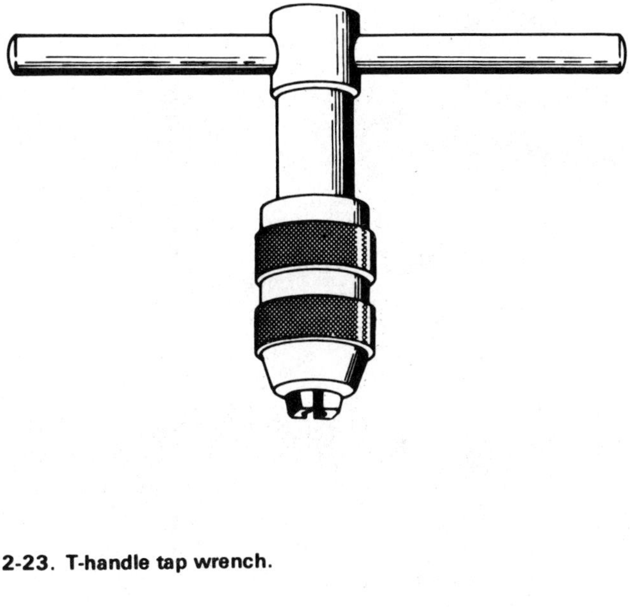

Fig. 2-23. T-handle tap wrench.

PIPE AND BOLT TAPS

A pipe or bolt tap differs from a pipe or bolt die in that a tap cuts an inside or female thread. Taps are used to cut new threads or to chase damaged threads in fittings, castings, etc. Fig. 2-24 shows some common taps.

TAP DRILL SIZES

When it is necessary to tape out a new thread, it is essential that the hole drilled for the tapping be the correct size. Table 2-1 shows the correct drill sizes for coarse and fine bolt threads and also for tapered pipe threads.

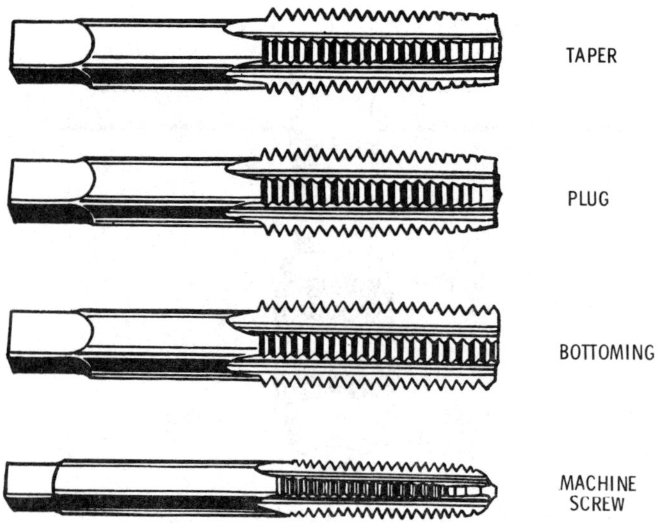

Fig. 2-24. Various kinds of taps.

PLIERS

A good pair of pliers (Fig. 2-25) is a very necessary tool for the plumber and pipe fitter. They should not be considered a substitute for a wrench, although pliers are often used as such, but rather an all-around utility tool.

Table 2-1. Tap and Drill Sizes Based on Approximately 75% Full Thread

National Coarse & Fine Threads

Thread	Drill	Thread	Drill	Taper Pipe	Drill
#0-80	3/64	#12-24	#17	1/8	R
#1-64	#53	#12-28	#15	1/4	7/16
#1-72	#53	1/4-20	#8	3/8	37/64
#2-56	#51	1/4-28	#3	1/2	23/32
#2-64	#50	5/16-18	F	3/4	59/64
#3-48	5/64	5/16-24	I	1	1 5/32
#3-56	#46	3/8-16	5/16	1 1/4	1 1/2
#4-40	#43	3/8-24	Q	1 1/2	1 47/64
#4-48	#42	7/16-14	U	2	2 7/32
#5-40	#39	7/16-20	W		
#5-44	#37	1/2-13	27/64		
#6-32	#36	1/2-20	29/64		
#6-40	#33	9/16-12	31/64		
#8-32	#29	9/16-18	33/64		
#8-36	#29	5/8-11	17/32		
#10-24	#25	5/8-18	37/64		
#10-32	#21				

Rod and bolt sizes are O.D. (outside measurement).
Pipe sizes are I.D. (inside measurement).

PIPE WRENCHES

The *straight pattern pipe wrench* (Fig. 2-26), available in six sizes from 6 in. through 60 in., is the basic tool of this trade. It is used to install and remove pipe and fittings.

The *end pattern pipe wrench* (Fig. 2-27) is used when working in close quarters or next to a wall or corner. Available in eight sizes, 6 in. through 36 in.

The *chain wrench* (Fig. 2-28), which can be used with a ratche-tlike action, makes it easier to work in extra close quarters. Made in four sizes from 14 in. through 36 in.

The *compound wrench* (Fig. 2-29) lets one man do the work of two. The short handle makes it easy to get at frozen joints in tight

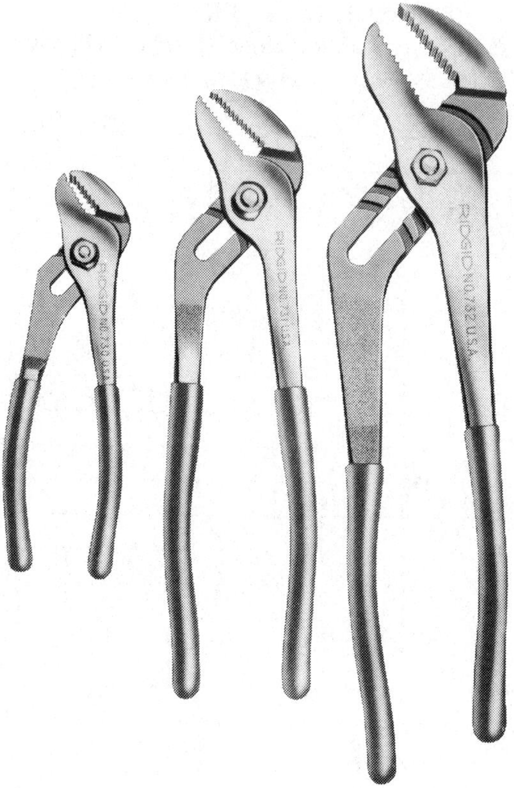

Courtesy Ridge Tool Co.

Fig. 2-25. Pliers.

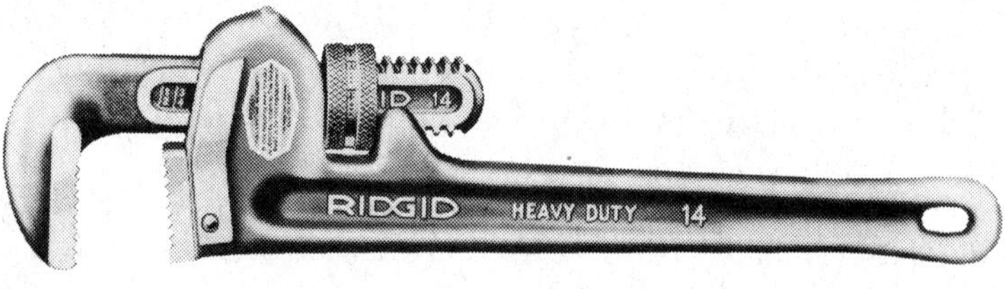

Courtesy Ridge Tool Co.

Fig. 2-26. A straight-pattern pipe wrench.

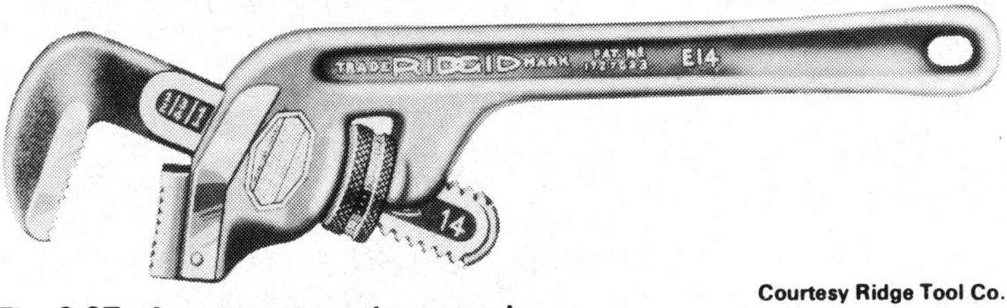

Courtesy Ridge Tool Co.

Fig. 2-27. An end-pattern pipe wrench.

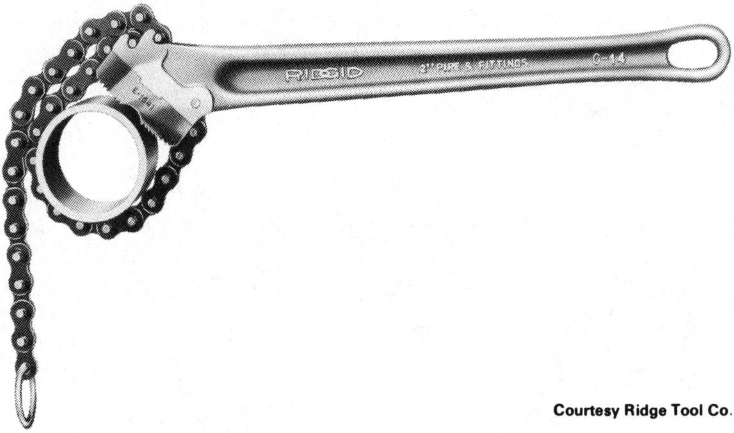

Courtesy Ridge Tool Co.

Fig. 2-28. A chain wrench.

Courtesy Ridge Tool Co.

Fig. 2-29. A compound wrench.

quarters. The turning force of the wrench is multiplied by compound leverage.

BASIN WRENCH

The basin wrench, shown in Fig. 2-30, is used primarily for loosening and tightening coupling nuts in hard-to-reach points behind sinks and lavatories where it is impossible to use a standard wrench. The jaws are made to swivel 180° for tightening or loosening action; the jaws are spring loaded to provide a ratcheting action. Some brands have a telescoping shank for adjustment to best working length.

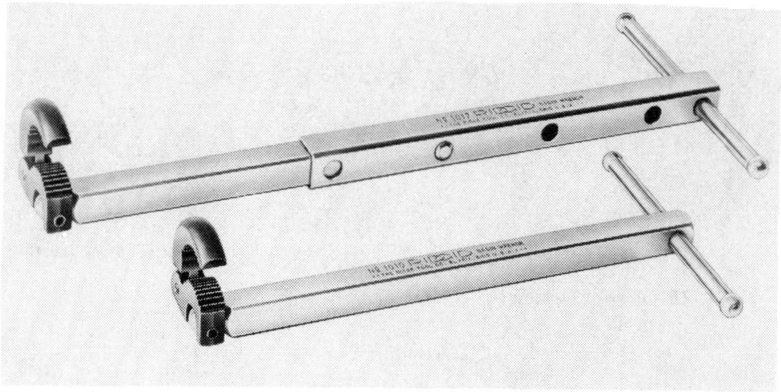

Courtesy Ridge Tool Co.

Fig. 2-30. A basin wrench.

ADJUSTABLE WRENCHES

The wrench shown in Fig. 2-31 has a thin head design for work in close quarters. Adjustable wrenches are indispensable tools in the plumbing and pipe fitting trades. They are especially useful for tightening nuts and bolts on flanged fittings and valves.

POWER TOOLS

In order to increase production and efficiency, power tools are used both in the shop and at job sites. When at locations where no electrical source is available, power is supplied by batteries or generators. Fig. 2-32 illustrates a typical battery-operated ¼-inch general-purpose drill. Attachments are available for power drills that convert them into buffers, jig saws, orbital sanders, right-angle drills, drill presses, bench sanders, bench grinders, and circular saws. Electric drills vary in type and size from the very small ¼-inch to the heavy-duty 1-inch chuck type.

Portable power saws are used for many purposes, both with hacksaw-type blades and with high-speed hole saws, as shown in Fig. 2-33 and 2-34. The hole saws are used with drill presses for

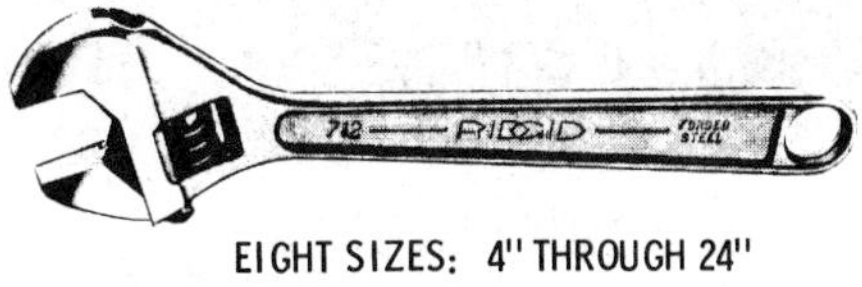

Fig. 2-31. An adjustable wrench.

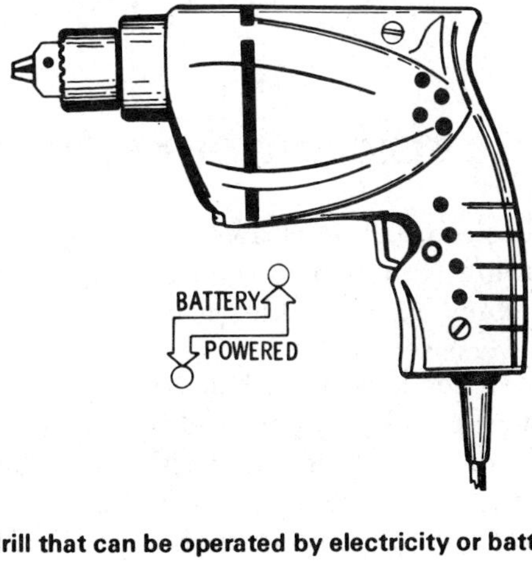

Fig. 2-32. A drill that can be operated by electricity or battery.

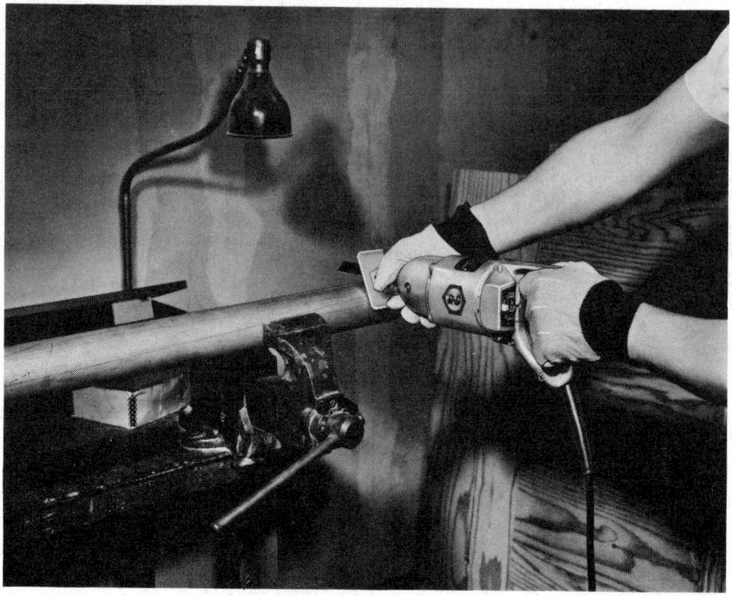

Courtesy Black & Decker

Fig. 2-33. A portable power hacksaw.

shop use and with portable drills on job sites. Blade selection, particularly for hacksaws, is governed by the type of material being cut.

PORTABLE ELECTRIC DRAIN CLEANERS

The portable electric drain cleaner shown in Fig. 2-35 is lightweight and has low starting torque for easy close-quarter operation. The power unit has a variable speed motor (0-500 rpm) with forward and reverse switch.

Recommended for lavatories, laundry tubs, sinks, bathtubs, closets, and shower stalls. Can be used with a 5/16-in. cable with bulb auger only, or with a 3/8-in. cable with either a bulb auger or a grease cutter. If you have only an occasional use for this tool, it can be rented (as can almost any other tool).

Courtesy Black & Decker

Fig. 2-34. A portable power drill using a high-speed hole saw.

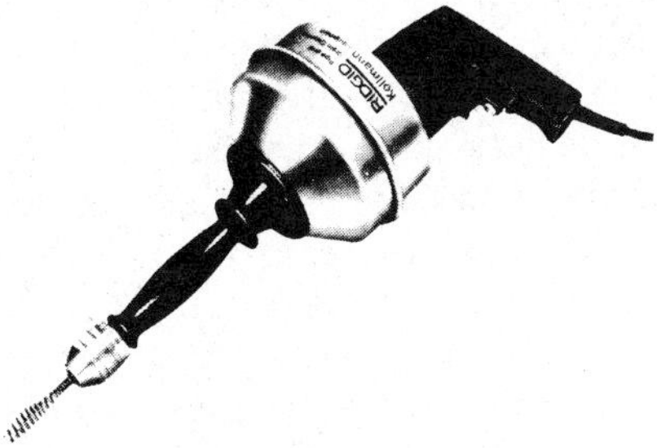

Fig. 2-35. A portable electric drain cleaner.

Table 2-2. Power Hacksaw Cutting Chart

Material	Teeth per Inch	Strokes per Minute	Weight or Pressure, Lbs.
Aluminum Alloy	4-6	150	60
Aluminum, Pure	4-6	150	60
Brass Castings, Soft	6-10	150	60
Brass Castings, Hard	6-10	135	60
Bronze Castings	6-10	135	125
Cast Iron	6-10	135	125
Copper, Drawn	6-10	135	125
*Carbon Tool Steel	6-10	90	125
*Cold Rolled Steel	4-6	135	150
*Drill Rod	10	90	125
*High Speed Steel	6-10	90	125
*Machinery Steel	4-6	135	150
Manganese Bronze	6-10	90	60
*Malleable Iron	6-10	90	125
*Nickel Silver	6-10	60	150
*Nickel Steel	6-10	90	150
Pipe, Iron	10-14	135	125
Slate	6-10	90	125
*Structural Steel	6-10	135	125
Tubing, Brass	14	135	60
*Tubing, Steel	14	135	60

*Use cutting compound or coolant.

SOLDERING

Soldering may be defined as the process of joining two metal parts together by a metal called *solder*. Solder is an alloy of lead and tin and has a lower melting point than either of its components or the metals to be joined. Solder that melt readily are termed *soft solders*, while those that melt at a red heat are termed *hard solders*. The process of soldering consists of cleaning the surfaces to be joined and heating them to the soldering temperature by any suitable means, such as a propane torch.

The essentials for any soldering job are clean metal surfaces, correct flux, good-quality solder, and sufficient heat.

The purpose of the soldering flux is to remove any grease or oxide present on the materials to be soldered. The solder is then melted into the joint and the joint smoothed over and finished by the use of a copper-tipped soldering iron or other heating means.

Torches and Lead-Melting Devices

A propane torch (Fig. 2-36) is often used for soldering, particularly when the work is performed on copper and brass pipe. Propane torches are easy to operate. The gas jet is turned on, a spark lighter or match is used, and immediately you have a 2250° flame. There are a number of attachments, such as a blowtorch burner head, flame spreader, or soldering tip, that can be readily obtained.

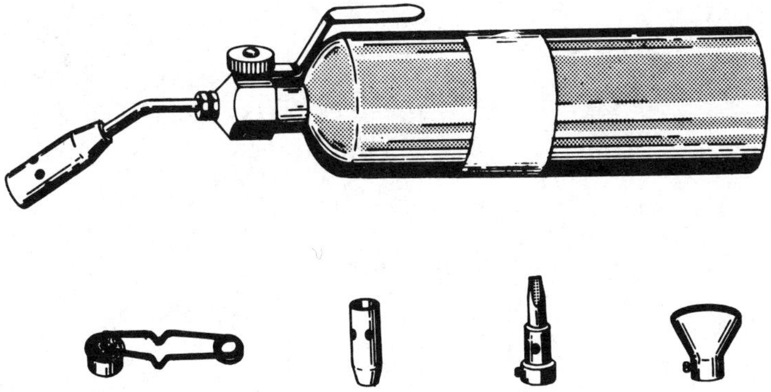

Fig. 2-36. A propane torch.

Fig. 2-37. Cast-iron melting pot and ladle.

For melting lead, a pot and ladle made of cast iron is used (Fig 2-37). The ladle is used for pouring lead into joints in cast-iron pipes. The best pattern of ladle is provided with three lips so that the melted material can be poured in any of three different directions.

CHAPTER 3

Pipe Fittings

Since pipe cannot be obtained in unlimited lengths, and in practically all pipe installations there are numerous changes in directions, branches, etc., pipe fittings have been devised for the necessary connections. By definition, the term *pipe fitting* is used to denote all fittings that may be attached to pipes:

1. To alter the direction of a pipe.
2. To connect a branch with a main.
3. To close an end.
4. To connect two pipes of different sizes.

There are many different types of pipe fittings. These various fittings may be classed,

1. With respect to material, as:
    a. Cast iron.
    b. Malleable iron.

    c. Brass.
    d. Copper.
    e. Steel (cast and forged).
    f. Plastic (nonmetallic).
    g. Glass.
2. With respect to design as:
    a. Plain.
    b. Beaded.
    c. Band.
3. With respect to the method of connecting, as:
    a. Screwed.
    b. Flanged.
    c. Hub-and-spigot.
    d. Cement.
    e. Soldered.
    f. Glued.
4. With respect to strength, as:
    a. Standard.
    b. Extra heavy.
    c. Double extra heavy.
5. With respect to the surface, as:
    a. Black.
    b. Galvanized.
6. With respect to finish, as:
    a. Rough.
    b. Semifinished.
    c. Polished.
7. With respect to service, as:
    a. Gas.
    b. Steam.
    c. Hydraulic (heavy pressure).
    d. Drainage.
    e. Railing.
    f. Sprinkler.
    g. Water.

The following definitions relating to pipes, joints, and fittings will be helpful to the pipe fitter and those desiring to acquire a knowledge of the subject.

DEFINITIONS

Bonnet—A cover used to guide and enclose the tail end of a valve spindle or a cap over the end of a pipe.

Branch—The outlet or inlet of a fitting not in line with the run, but which may make any angle.

Branch Ell—Used to designate an elbow having a back outlet in line with one of the outlets of the "run." It is also called a heel-outlet elbow, and is incorrectly used to designate a side-outlet or back-outlet elbow.

Branch Pipe—A very general term used to signify a pipe equipped with one or more branches. Such pipes are used so frequently that they have acquired common names such as tees, crosses, side- or back-outlet elbows, manifolds, double-branch elbows, etc. The term *branch pipe* is generally restricted to those pipes that do not conform to usual dimensions.

Branch Tee—A tee having many side branches. *See* Manifold.

Bull-head Tee—A tee, the branch of which is larger than the run.

Bushing—A pipe fitting for the purpose of connecting a pipe with a fitting of a larger size, being a hollow plug with internal and external threads to suit the different diameters.

Close Nipple—A nipple that is about twice the length of a standard pipe thread and is without any shoulder.

Coupling—A threaded sleeve used to connect two pipes. Commercial couplings are threaded to suit the exterior thread of the pipe. The term *coupling* is occasionally used to mean any jointing device and may be applied to either straight or reducing sizes.

Cross—A pipe fitting with four branches arranged in pairs, each pair on one axis, and the axes at right angles. When the outlets are arranged otherwise the fittings are branch pipes or specials.

Crossover—A fitting with a double offset, or shaped like the letter U with the ends turned out. It is made only in small sizes and is used to pass the flow of one pipe past another when the pipes are in the same plane.

Crossover Tee—A fitting made along the lines similar to a crossover, but having at one end two openings in a tee head, the plane of which is at right angles to the plane of the cross over bend.

Cross Valve—A valve fitted on a transverse pipe so as to open communication at will between two parallel lines of piping.

Crotch—A fitting that has a general shape of the letter Y. Caution should be exercised not to confuse the crotch and wye (Y).

Drop Ell (or *Wing Ell*)—A 90° ell with extensions cast on side for anchoring to wall surface.

Drop Tee—One having the same peculiar wings as the drop elbow.

Dry Joint—A joint made without gasket, packing or compound of any kind, as a ground joint.

Elbow (ell)—A fitting that makes an angle between adjacent pipes. The angle is always 90° unless another angle is stated. *See* Branch, Service and Union Ell.

Extra Heavy—When applied to pipe, means pipe thicker than standard pipe; when applied to valves and fittings, indicates units suitable for a working pressure of 250 lbs. per sq. in.

Header—A large pipe into which one set of boilers is connected by suitable nozzles or tees, or similar large pipes from which a number of smaller ones lead to consuming points. Headers are often used for other purposes—for heaters or in refrigeration work. Headers are essentially branch pipes with many outlets, which are usually parallel. Largely used for tubes or water-tube boiler.

Hub-and-Spigot Joint—The usual term for the joint in cast-iron pipe. Each piece is made with an enlarged diameter or hub at one end into which the plain or spigot end of another piece is inserted when laying. The joint is then made tight by cement, oakum, lead, rubber, or other suitable substance, which is driven in or calked into the hub and around the spigot. Applied to fittings or valves, the term means that one end of the run is a "hub" and the other end is a "spigot," similar to those used on regular cast-iron pipe.

Hydrostatic Joint—Used in large water mains, and in which sheet lead is forced tightly into the bell of a pipe by means of the hydrostatic pressure of a liquid.

Lead Joint—Generally used to signify the connection between pipes made by pouring molten lead into the annular space between a bell and spigot and then making the lead tight by calking.

Lead Wool—A material used in place of molten lead for making pipe joints. It is lead fiber, about as coarse as fine excelsior, and when made in a strand, it can be calked into the joints, making them very solid.

Lip Union—A special form of union characterized by a lip that prevents a gasket from being squeezed into the pipe to obstruct the flow. It is a ring union unless flange is specified.

Manifold—A fitting with numerous branches used to convey fluids between a large pipe and several smaller pipes. *See* Branch Tee. A header for a coil.

Medium Pressure—When applied to valves and fittings, means suitable for a working pressure of from 125 to 175 lbs. per sq. in.

Needle Valve—A valve provided with a long tapering point in place of the ordinary valve disk. The tapering point permits fine graduation of the opening. At times called a *needle-point* valve.

Nipple—A short piece of pipe, threaded on both ends, can be any length from a close nipple up to and including a 12-in. nipple. Pipe over 12 in. is regarded as cut pipe.

Reducer—A fitting with a female pipe thread on each end, one thread is one or more pipe sizes smaller than the other, it is technically a reducing coupling. The largest size is always correctly named first, as: 1 ½″ × 1 ¼″ reducer. While other fittings, ells, tees, and wyes can be used to reduce pipe sizes, the term reducer is correctly applied only to a reducing coupling.

Roof increaser—A fitting designed to increase the size of a small waste or bent stack at the point where it exits the building at the roof line. *See* Fig. 3-25G.

Run—A length of pipe made up of more than one piece of pipe. The portion of any fitting having its end "in line" or nearly so, in contrast to the branch or side opening, as of a tee.

Service Pipe—A pipe connecting mains with a building.

Shoulder Nipple—The next length to a close nipple, the actual amount of unthreaded pipe between the two threads will vary with the pipe size.

Standard Pressure—A term applied to valves and fittings suitable for a working steam pressure of 125 lbs. per sq. in.

Street Elbow—An elbow having one male thread and one female thread.

Street Tee—A tee having one male thread (end) and two female threads (one end and side).

Tee—A fitting that has one side outlet at right angles to the run. A bullhead tee is one in which the side outlet is larger than the run.

Union—A fitting used to connect pipes. There are two common types of unions; the flange union (Fig. 3-17) and the three piece union (Fig. 3-8).

Union Ell—Union Ells are a combination union and elbow, in one fitting. They are made in both male and female patterns, and in 90° and 45° ells. (They are shown in Fig. 3-18.)

Union Tee—Union Tees are a combination union and tee in one fitting. They are made in both run and outlet patterns as shown in Fig. 3-18. Union tees are not a commonly used fitting.

Wye—A fitting that has a side outlet at an angle other than 90°. The angle would be considered to be 45° unless another angle is specified. Wyes are made in single or double patterns.

CAST-IRON FITTINGS

Standard beaded or flat-band fittings of cast iron are suitable for 125 lbs. or steam or 175 lbs. of water pressure. These fittings will actually require from 1000 to 2500 lbs. of pressure to burst them. The larger factor of safety is necessary in their use because of the strain due to expansion, contraction, weight of piping, settling, and water hammer. For steam pressures above 125 lbs., extra-heavy xfittings should be used.

MALLEABLE-IRON FITTINGS

Standard beaded or flat-band fittings of malleable iron are intended for working steam pressure up to 150 lbs. Such fittings have, at various times, been subjected to hydraulic pressures of from 2000 to 4000 lbs. without bursting. It would therefore seem possible that they would be safe for at least 250 lbs. of working steam pressure. If proper care is exercised in fitting and using them, they will undoubtedly be found satisfactory for working pressure up to 500 lbs. However, since all fittings are subjected to strain due to expansion, contraction, and making up the joints, they are not recommended for working pressures over 150 lbs. In fact, since extra-heavy fittings cost only a little more, it is not economical to use standard fittings for working pressures near 150 lbs. Standard plain-pattern malleable fittings are used for low-pressure gas and water and house plumbing.

BRASS FITTINGS

Brass fittings are made in standard, extra-heavy, and cast-iron patterns (iron-pipe sizes), and are used for brass feed-water pipes where hard water makes steel pipes undesirable. The standard brass fittings are usually made in sizes from 1/4 to 3 in., suitable for 125 lbs. working pressure; extra-heavy fittings, 1/8 to 6 in., suitable for 150 lbs. working pressure; cast-iron patterns in all sizes, suitable for 250 lbs. working pressure.

CAST-STEEL FITTINGS

Cast-steel fittings are made extra heavy with screwed or flanged ends. The screwed fittings are listed in sizes from 3 to 6 in. The 3- to 4 1/2 in. sizes (inclusive) are tested for 1500 lbs. hydrostatic pressure, and the 5- and 6-in. sizes for 1200 lbs. pressure. The radii of these fittings are larger than ordinary fittings, thereby reducing friction. They are suitable for the working pressures given when used in hydraulic installations in which shock is absent or so slight as to be negligible.

Ordinarily, these fittings, when subject to shock, should not be used for working pressures higher than 65 percent of the hydrostatic test pressure, and where shock is severe, 50 percent, or even 40 percent, will be more conservative. Installations of this character should always be protected by shock absorbers placed to the best advantage.

FORGED-STEEL FITTINGS

The extra-heavy hydraulic forged-steel screwed fittings are suitable for superheated steam up to 2350 lbs. working pressure, a total temperature of 899° F, and for cold water or oil working hydrostatic pressures up to 3000 lbs. They are regularly made from solid forgings in sizes ranging from ½ to 2½ in. inclusive, and are tested to 3000 lbs. hydraulic pressure. The double extra-heavy pattern is suitable for cold water or oil working hydrostatic pressures up to 6000 lbs. They are regularly made from solid forgings in sizes ranging from ⅜ to 2 in. inclusive, and are tested to 6000 lbs. hydrostatic pressure.

VARIOUS FITTINGS

There are a great many fittings due to the many modifications of each class of fittings and the several weights and different metals used. A list of these fittings may be divided into several groups, classified with respect to their use:

1. Extension or joining.
    a. Nipples.
    b. Lock nuts.
    c. Couplings.
    d. Offsets.
    e. Joints.
    f. Unions.
2. Reducing or enlarging.
    a. Bushings.
    b. Reducers.
3. Directional.
    a. Offsets.
    b. Elbows.
    c. Return bends.
4. Branching.
    a. Side-outlet elbows.
    b. Back-outlet return bends.
    c. Tees.

d. Y branches.
e. Crosses.
5. Shut off or closing.
a. Plugs.
b. Caps.
c. Flanges.
6. Union
a. Union elbows.
b. Union tees.

Nipples

By definition, a nipple is a *piece of pipe less than 12 in. in length threaded on both ends*; pipe over 12 in. long is regarded as cut pipe. With respect to length, nipples may be classed as:

1. Close.
2. Short.
3. Long.

Where fittings or valves are to be very close to each other, the intervening nipple is just long enough to take the threads at each end, being called a *close nipple*. If a small amount of pipe exists between the threads, it is called a *shoulder* or *short nipple*, and where a larger amount of bare pipe exists, it is called a *long nipple* or *extra-long nipple*.

Nipples having a right-hand thread on one end and a left-hand thread on the other are generally used in steam heating piping instead of unions. Fig. 3-1 shows such a nipple, with a hexagon nut at the center forming part of the nipple.

Lock Nuts

Lock nuts are made with faced end for use on long screw nipples having couplings, and with a recessed or grooved end to hold

Fig. 3-1. A right- and left-hand threaded nipple with hexagon nut in the center of the nipple.

packing where this is depended on to make a tight joint. The use of lock nipples should be avoided wherever possible as the joint is not as good as that obtained by a union.

Couplings

An ordinary coupling, shown in Fig. 3-2, usually comes with the pipe, one coupling to each length. The couplings are made of wrought or cast metal, or of brass. They are regularly threaded with right-hand threads, but can be obtained on special order with right- and left-hand threads. R and L couplings have projecting bars or rings to distinguish them from standard couplings. Another form of coupling, called an *extension piece*, is shown in Fig. 3-3; it differs from the standard coupling in that it has a male thread at one end. There are numerous other types, some known as *reducers* (Fig. 3-4) and others as joints. The most important of these are the screw or threaded joint (already described), and the flanged joint. Fig. 3-5 shows various screwed joints.

Extension Coupling—An extension coupling (Fig. 3-3) is a fitting with a male pipe thread on one end and a female pipe thread

Fig. 3-2. A pipe coupling.

Fig. 3-3. A pipe extension piece.

Fig. 3-4. A pipe reducer.

on the other. It is used primarily for extending existing piping opening to a new wall line; for instance, if ceramic tile is added to an existing wall, an extension coupling, added to the existing fitting will extend the existing piping to the same relative distance behind the new wall line.

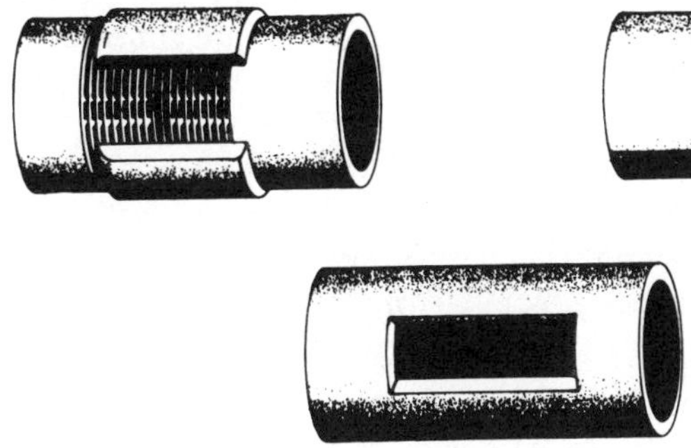

Fig. 3-5. Various screw pipe couplings.

Unions

There are various kinds of unions. The plain union, as shown in Fig. 3-6, requires a gasket; the two pipes to be joined by the union must be in approximate alignment to secure a tight joint because of

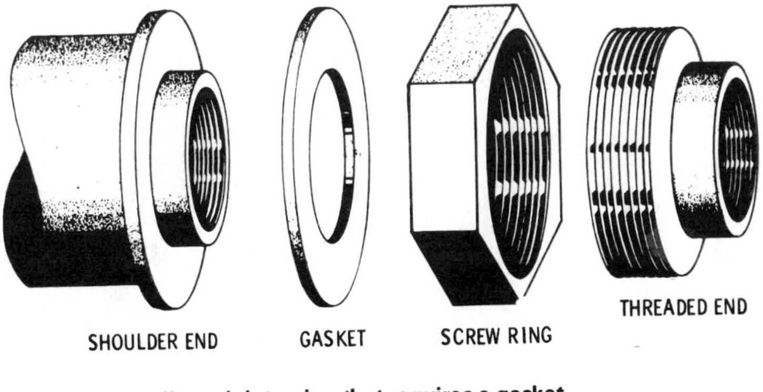

Fig. 3-6. An ordinary joint union that requires a gasket.

the flat surfaces that must press against the gasket. This limitation is shown in Fig. 3-7.

To avoid this difficulty, and also to avoid the inconvenience of the gasket, various unions having spherical seats and ground joints have been devised. These consist of a composition ring bearing against iron, or with both contact surfaces of composition. Fig. 3-8 shows the construction of a *ground-joint union*. The joint has spherical contact, and the illustration shows the tight joint secured, even though the pipes are out of line. Unions are also made entirely of brass with ground joints.

Right and Left Nipples—Right and left nipples are made to serve as unions, primarily in steam coils. Right and left nipples must be screwed into a right and left fitting. Right and left nipples have largely been replaced by newer methods of connection.

Bushings

Bushings are often confused with reducers. The function of a bushing is to *connect the male end of a pipe to a fitting of larger*

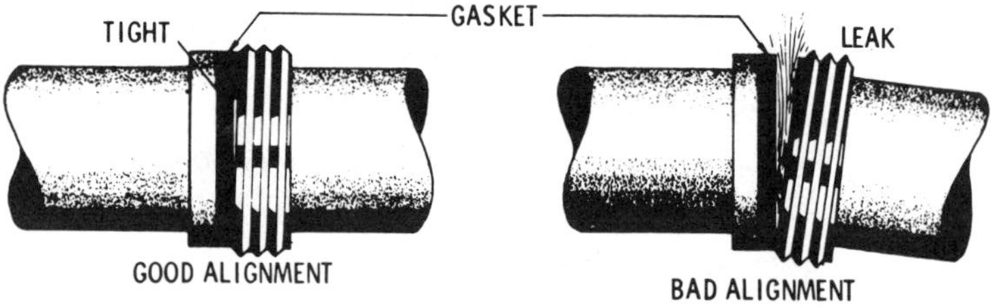

Fig. 3-7. The importance of aligning pipes.

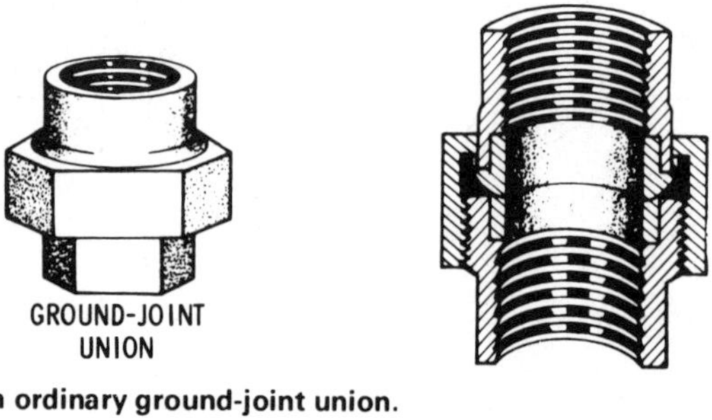

Fig. 3-8. An ordinary ground-joint union.

size. It consists of a hollow plug with male and female threads to suit the different diameters. A bushing may be regarded as either a reducing or an enlarging fitting. As generally manufactured, bushings 2½ in. and smaller, reducing one size, are malleable iron; reducing two or more sizes are cast iron; all above 2½ in. are cast iron, except brass bushings, which may be obtained in sizes from ¼ to 4 in.

Bushings are listed by the *pipe size of the male thread;* thus a ¼ in. bushing joins a ¼ in. fitting to a ⅛ in. pipe. To avoid mistakes, however, it is better to specify the size of both threads, for instance, calling the bushing just mentioned a ¼″ × ⅛″ bushing. The regular-pattern bushing has a hexagon nut at the female end for screwing the bushing into the fitting. The *faced* bushing is used for very close work, having a faced end in place of the hexagon nut. This may be used with a long screw pipe and faced lock nut to form a tight joint or to receive a male end fitting for close work. Fig. 3-9 shows the plain and faced types of bushing. As shown in Fig. 3-9 bushings are made in several different patterns: Concentric Hex and Face bushings and Eccentric Hex Bushings.

Concentric bushings are used when it is necessary to keep both sizes of piping on the same center.

Eccentric bushings are used when it is desirable to keep the top plane of both sizes of piping as near level as possible. At times, for instance in a hot-water heating system, it is important that both sizes of piping be on the same plane in order to eliminate air problems.

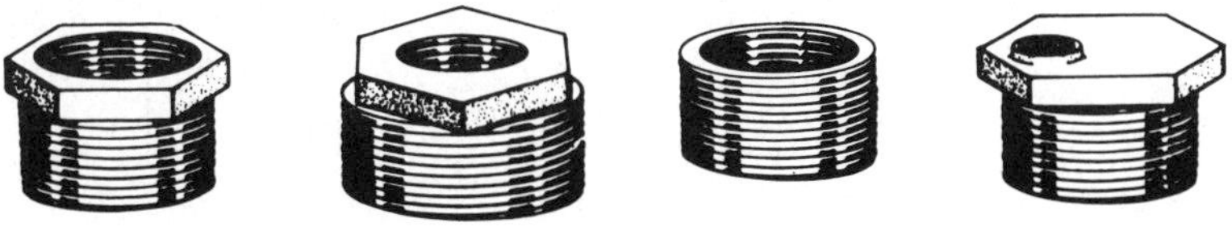

Fig. 3-9. Various bushing.

Return Bends

Return bends are largely used for making up pipe coils for steam heating and for water boilers. They are U-shaped fittings with a female thread at both ends and are regularly made in three patterns, known as:

1. Close.
2. Medium.
3. Open.

Some manufacturers also make an extra-close and an extra-wide pattern. These patterns represent various widths between the two arms. There seems to be no standard as to the spacing of the arms for the different patterns; hence, for close work, the fitter should ascertain the center-to-center dimensions from the manufacturer's catalogue of the make to be used. Table 3-1 gives the dimensions of the various elbows. Return bends smaller than the 1/2-in. size are difficult to buy since very few manufacturers make them. It should be noted that return bends are made in 1/4 in. and 3/8 in. sizes.

Table 3-1. Malleable Return Bends

	Extra Close		Close		Medium		Open		Wide
Size	Center to Center	Weight per 100 (plain)	Center to Center	Weight per 100 (banded)	Center to Center	Weight per 100 (banded)	Center to Center	Weight per 100	Center to Center
1/4	....	....	3/4	15.5	....	....	1 1/8	21	....
3/8	....	....	7/8	22	....	....	1 1/4	22	....
1/2	....	....	1 1/8	35	1 1/4	34	1 1/2	44	....
3/4	1 1/4	77	1 3/8	71	1 9/16	60	2	83	6
1	1 9/16	92	1 3/4	100	1 7/8	92	2 1/2	140	3, 4, 4 1/2 5, 6, 7, 8
1 1/4	....	....	2 1/8	168	2 1/4	160	3	200	4, 5, 6, 9
1 1/2	....	....	2 1/2	244	2 9/16	255	3 1/2	310	5, 6, 8
2	....	....	2 3/4	388	3 1/16	337	4 3/8	550	2, 6, 7, 8
2 1/2	....	....	3 7/8	631	....	....	4 3/4	710	....
3	....	....	4 1/2	880	....	....	6 1/4	1050	....
3 1/2	....	....	5	1400	....	....	6 1/2	1550	....
4	....	....	....	....	....	....	7	1850	....
5	....	....	6	....	....	....	....	....	....

For making up so-called coils from short lengths of pipe, return bends may be obtained with elbows, as illustrated in Fig. 3-10. The pipes, when screwed into the fitting, will not be parallel but will spread like the sides of the letter V. Such bends are usually listed for which the pitch is suitable.

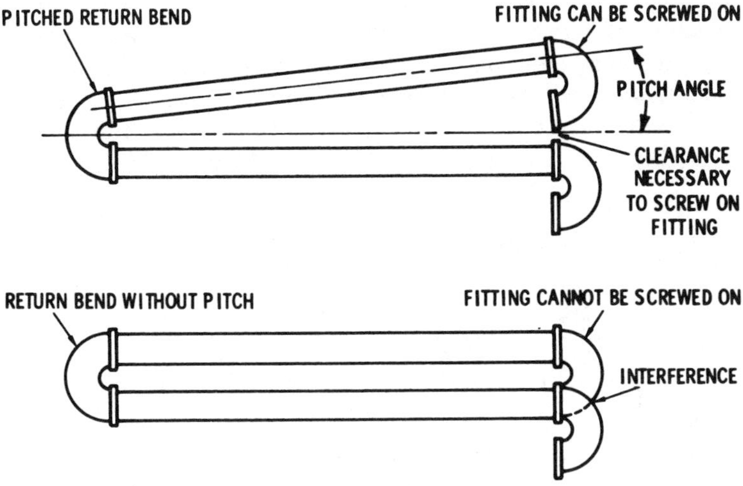

Fig. 3-10. Coils made with short pipe lengths and various angle elbows.

Side-Outlet Elbows

The two openings of an elbow indicate its *run*, and when there is a third opening, the axis of which is at 90° to the plane of the run, the fitting is a side-outlet elbow, as shown in Fig. 3-11. These fittings are regularly made in sizes ranging from ¼ in. to 2 in. inclusive, with all outlets of equal size and with the side outlet one and two sizes smaller than the rim outlets. In general, it is not well to specify fittings of this kind, which are not in much demand, unless the more usual forms are difficult to find.

Back- and Side-Outlet Return Bends

These types of bends are simply return bends provided with an additional outlet at the back or side, as shown in Fig. 3-12. They are regularly made in sizes ranging from ¾ in. to 3 in. inclusive, in the close or open patterns.

Tees

Tees are the most important and widely used of the branching fittings. Tees, like elbows, are made in many different sizes and patterns. They are used for making a branch of 90° to the main pipe, and always have the branch at right angles. When the three outlets are of the same size, the fitting is specified by the size of the pipe, as a ½ in. tee; when the branch is a different size than the run outlets, the size of the run is given first, as a 1″ × ¼″ tee. When all three outlets are of different sizes, they are all specified, giving the sizes of the run first as a 1¼″ × 1″ × 1½″ tee.

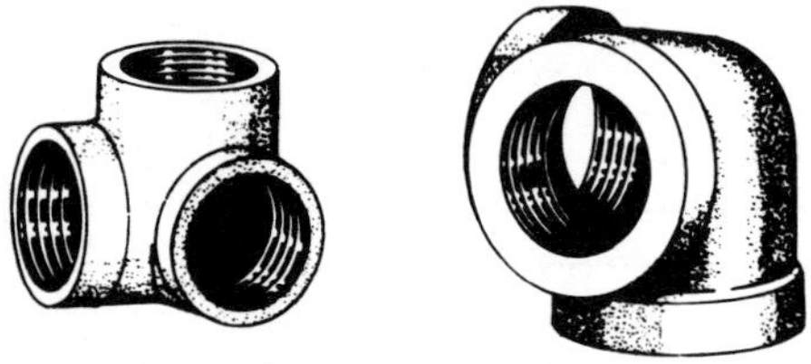

Fig. 3-11. Cast-iron elbows with side outlets.

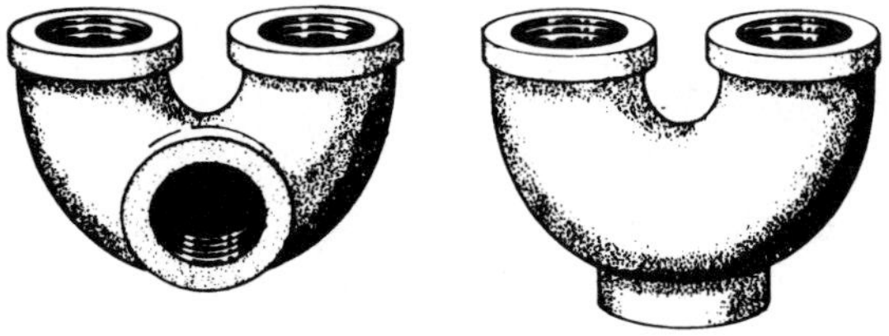

Fig. 3-12. Cast-iron return bends and side outlets.

Wyes

A wye is a fitting with the side opening set at an angle; 45° wyes are the most commonly used. Wyes are made in both single and double patterns, and are made in straight and reducing sizes. The correct way to "read" a wye is: end, end, and side. Thus a wye with all openings being 1½ in. is a 1½-in. wye; a wye with 1½-inch end openings and a 1¼-inch side opening is a 1½″ × 1½″ × 1¼″ wye.

Crosses

A cross is simply an ordinary tee having a back outlet opposite the branch outlet (Fig. 3-13). The axes of the four outlets are in the same plane and at right angles to each other. Crosses, like tees, are made in a number of sizes. Regarding a cross as a tee with a back outlet, the tee part is made in various combinations of sizes, similar to ordinary tees, but the back outlet is always the same size as the opposite side outlet of the tee parts.

Plugs

A *plug* is used for closing the end of a pipe or a fitting having a female thread. In metal, plugs are made of cast iron, malleable iron, and brass. Fig. 3-14 shows the various patterns. Usually a square head or a four-sided countersunk head is used for the small sizes, and a hexagon head for the larger sizes. Ordinary metal plugs are made in sizes ranging from ⅛ in. to 12 in. inclusive.

Caps

A *cap* is used for closing the end of a pipe or fitting having a male thread. Caps, like plugs, are made of cast iron, malleable iron, and brass. Fig. 3-15 shows various cap designs. Plain and flat-band or beaded caps are regularly made in sizes from ⅛ in. to 6 in. inclusive; cast-iron caps from ⅜ in. to 15 in. inclusive.

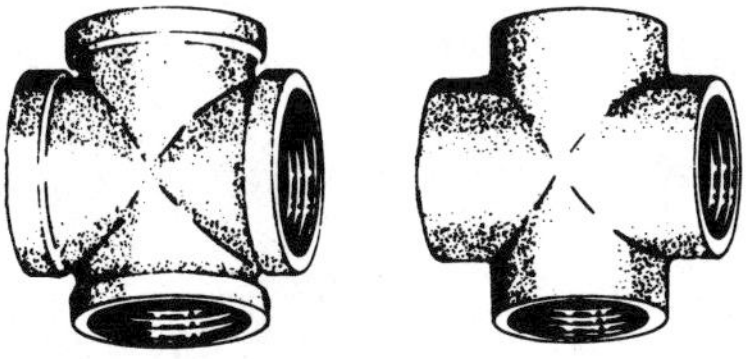

Fig. 3-13. Two cross connectors.

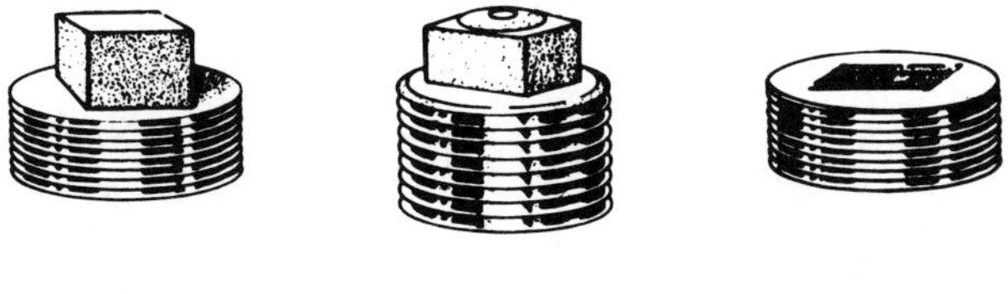

Fig. 3-14. Various pipe plugs.

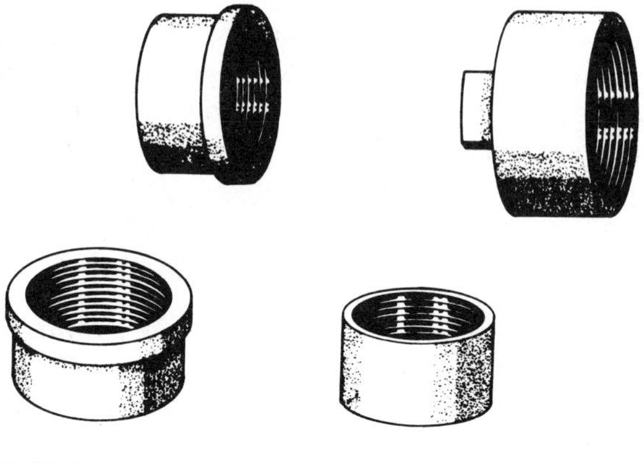

Fig. 3-15. Various pipe caps.

Flanges

Flanges are made of copper and brass for use with copper piping and of steel for use with steel or cast-iron piping and equipment (Fig. 3-16). Flanges are made for use with threaded piping, for welded piping (slip-on and weld neck types), and solder type, for soft or hard soldering. Flanges are made in straight pipe sizes for use when bolting piping and equipment together, and are also made in reducing sizes. They are also made for closing the end of a run of piping; this type is called a *blind flange.*

Flanged Fittings and Valves—Flanged fittings and valves are commonly used on piping 2½ in. and larger (Fig. 3-17). The use of flanged fittings and valves permits easy removal of pumps, specialized equipment, etc. Flanged fittings and valves are used on

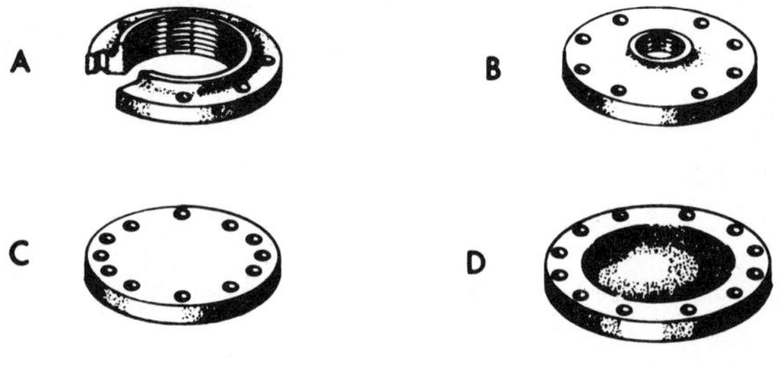

Fig. 3-16. Various flanges: (A) screwed flange; (B) screwed reducing flange; (C) and (D) blind flanges.

Table 3-2. Sizes and Dimensions of Typical Metal Flanges

Pipe Size and Diameter (inches)	Thickness (inches): Through Flange — Flange A	Thickness (inches): Through Hub — Flange A	Thickness (inches): Through Flange — Flanges C & D	Bolt sizes: Number of Holes	Bolt sizes: Dia. of Bolts (inches)
1 × 4 1/4	7/16	11/16	7/16	4	1/2
1 1/4 × 4 5/8	1/2	13/16	1/2	4	1/2
1 1/2 × 5	9/16	7/8	9/16	4	1/2
2 × 6	5/8	1	5/8	4	5/8
2 1/2 × 7	11/16	1 3/16	11/16	4	5/8
3 × 7 1/2	3/4	1 1/4	3/4	4	5/8
3 1/2 × 8 1/2	13/16	1 1/4	13/16	8	5/8
4 × 9	15/16	1 5/16	15/16	8	5/8
5 × 10	15/16	1 7/16	15/16	8	3/4
6 × 11	1	1 9/16	1	8	3/4
8 × 13 1/2	1 1/8	1 3/4	1 1/8	8	3/4
10 × 16	1 3/16	1 15/16	1 3/16	12	7/8
12 × 19	1 1/4	2 3/16	1 1/4	12	7/8

Flange B

Pipe Size and Dia. (inches)	Thickness Through Flange (in.)	Bolt Sizes: Number of Holes	Bolt Sizes: Dia. of Bolts	Pipe Size and Dia. (inches)	Thickness Through Flange (in.)	Bolt Sizes: Number of Holes	Bolt Sizes: Dia. of Bolts
1 × 5	9/16	4	1/2	3 × 10	15/16	8	3/4
1 × 6	5/8	4	5/8	4 × 10	16/16	8	3/4
1 1/4 × 6	5/8	4	5/8	2 × 11	1	8	3/4
1 1/2 × 6	5/8	4	5/8	2 1/2 × 11	1	8	3/4
1 1/2 × 7	11/16	4	5/8	3 × 11	1	8	3/4
2 × 7	11/16	4	5/8	4 × 11	1	8	3/4
1 1/2 × 7 1/2	3/4	4	5/8	5 × 11	1	8	3/4
2 × 7 1/2	3/4	4	5/8	3 × 13 1/2	1 1/8	8	3/4
2 1/2 × 7 1/2	3/4	4	5/8	4 × 13 1/2	1 1/8	8	3/4
3 × 8 1/2	13/16	8	5/8	5 × 13 1/2	1 1/8	8	3/4
1 1/2 × 9	15/16	8	5/8	6 × 13 1/2	1 1/8	8	3/4
2 × 9	15/16	8	5/8	6 × 16	1 3/16	12	7/8
2 1/2 × 9	15/16	8	5/8	8 × 16	1 3/16	12	7/8
3 × 9	15/16	8	5/8	8 × 19	1 1/4	12	7/8
3 1/2 × 9	15/16	8	5/8	10 × 19	1 1/4	12	7/8

water, gas, steam, hot water, air, air conditioning, and sewage piping, as well as other specialized piping. Flanges used with these fittings are called *companion* flanges because they are made in standard sizes to mate with standard valves and fittings.

Table 3-3.
Templates for Bronze and Corrosion-Resistant Flanges

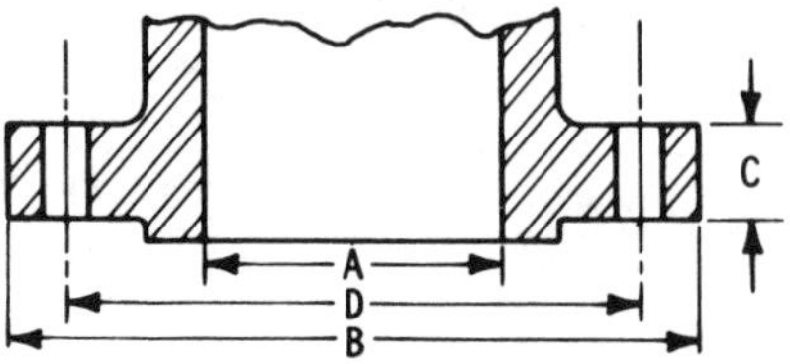

Dimensions, in Inches

Size A	B	C Bronze	C Corrosion-Resistant	D	Number of bolts	Diameter of bolts	Length of bolts	Length of stud bolts with 2 nuts
150-lb. Bronze ASA or Corrosion-Resistant MSS Standards								
1/2	3 1/2	5/16	3/8	2 3/8	4	1/2	1 1/4	1 7/8
3/4	3 7/8	11/32	13/32	2 3/4	4	1/2	1 1/2	1 7/8
1	4 1/4	3/8	7/16	3 1/8	4	1/2	1 1/2	2
1 1/4	4 5/8	13/32	1/2	3 1/2	4	1/2	1 1/2	2
1 1/2	5	7/16	9/16	3 7/8	4	1/2	1 1/2	2 1/8
2	6	1/2	5/8	4 3/4	4	5/8	1 3/4	2 1/2
2 1/2	7	9/16	5/8	5 1/2	4	5/8	2	2 5/8
3	7 1/2	5/8	5/8	6	4	5/8	2	2 3/4
*3 1/2	8 1/2	11/16	...	7	8	5/8	2 1/4	2 7/8
4	9	11/16	11/16	7 1/2	8	5/8	2 1/4	2 7/8
*5	10	3/4	...	8 1/2	8	3/4	2 1/2	3 1/4
6	11	13/16	13/16	9 1/2	8	3/4	2 1/2	3 3/8
8	13 1/2	15/16	15/16	11 3/4	8	3/4	2 3/4	3 5/8
10	16	1	1	14 1/4	12	7/8	3 1/4	4 1/8
12	19	1 1/16	1 1/16	17	12	7/8	3 1/4	4 1/4
300-lb. Bronze ASA Standard								
1/2	3 3/4	1/2	...	2 5/8	4	1/2	1 3/4	2 1/4
3/4	4 5/8	17/32	...	3 1/4	4	5/8	2	2 1/2
1	4 7/8	19/32	...	3 1/2	4	5/8	2	2 5/8
1 1/4	5 1/4	5/8	...	3 7/8	4	5/8	2	2 3/4
1 1/2	6 1/8	11/16	...	4 1/2	4	5/8	2 1/4	3 1/8
2	6 1/2	3/4	...	5	8	3/4	2 1/4	3
2 1/2	7 1/2	13/16	...	5 7/8	8	3/4	2 1/2	3 3/8
3	8 1/4	29/32	...	6 5/8	8	3/4	2 3/4	3 1/2
3 1/2	9	31/32	...	7 1/4	8	3/4	3	3 5/8
4	10	1 1/16	...	7 7/8	8	3/4	3	3 7/8
5	11	1 1/8	...	9 1/4	8	3/4	3 1/4	4
6	12 1/2	1 3/16	...	10 5/8	12	3/4	3 1/4	4 1/8
8	15	1 3/8	...	13	12	7/8	3 3/4	4 3/4

Table 3-4. Templates for 125-lb. Cast-Iron Flanges

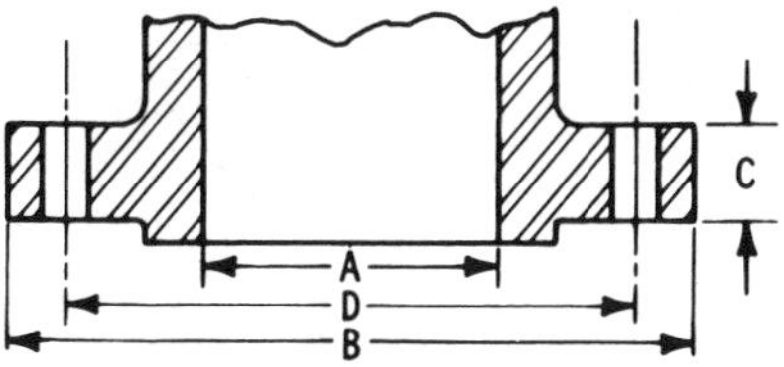

Size A	B	C	D	Number of bolts	Diameter of bolts	Length of bolts	Length of stud bolts with 2 nuts
125-lb. Cast Iron							
1	4 1/4	7/16	3 1/8	4	1/2	1 3/4	. . .
1 1/4	4 5/8	1/2	3 1/2	4	1/2	2	. . .
1 1/2	5	9/16	3 7/8	4	1/2	2	. . .
2	6	5/8	4 3/4	4	5/8	2 1/4	. . .
2 1/2	7	11/16	5 1/2	4	5/8	2 1/2	. . .
3	7 1/2	3/4	6	4	5/8	2 1/2	. . .
3 1/2	8 1/2	13/16	7	8	5/8	2 3/4	. . .
4	9	15/16	7 1/2	8	5/8	3	. . .
5	10	15/16	8 1/2	8	3/4	3	. . .
6	11	1	9 1/2	8	3/4	3 1/4	. . .
8	13 1/2	1 1/8	11 3/4	8	3/4	3 1/2	. . .
10	16	1 3/16	14 1/4	12	7/8	3 3/4	. . .
12	19	1 1/4	17	12	7/8	3 3/4	. . .
14	21	1 3/8	18 3/4	12	1	4 1/4	. . .
16	23 1/2	1 7/16	21 1/4	16	1	4 1/2	. . .
18	25	1 9/16	22 3/4	16	1 1/8	4 3/4	. . .
20	27 1/2	1 11/16	25	20	1 1/8	5	. . .
24	32	1 7/8	29 1/2	20	1 1/4	5 1/2	. . .
30	38 3/4	2 1/8	36	28	1 1/4	6 1/4	. . .
36	46	2 3/8	42 3/4	32	1 1/2	7	. . .
42	53	2 5/8	49 1/2	36	1 1/2	7 1/2	. . .
48	59 1/2	2 3/4	56	44	1 1/2	7 3/4	. . .
54	66 1/4	3	62 3/4	44	1 3/4	. . .	10 1/2
60	73	3 1/8	69 1/4	52	1 3/4	. . .	10 3/4
72	86 1/2	3 1/2	82 1/2	60	1 3/4	. . .	11 1/2
84	99 3/4	3 7/8	95 1/2	64	2	. . .	12 3/4
96	113 1/4	4 1/4	108 1/2	68	2 1/4	. . .	14

Table 3-5. Templates for 150- and 300-lb. Steel Flanges

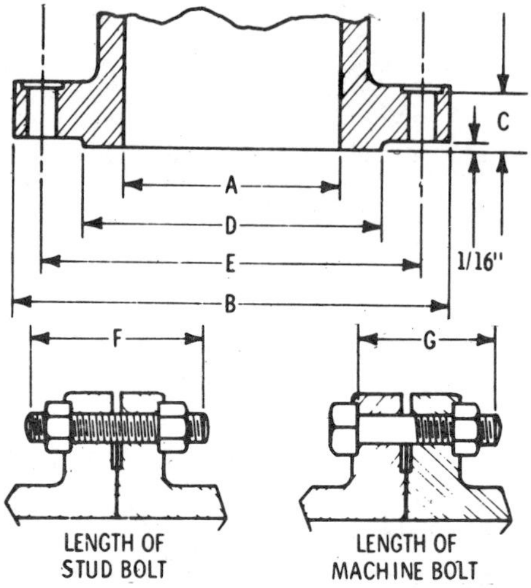

Size	A Valve or fitting	B	C Companion flange	C Valve or fitting	D	E	Bolts or Stud Bolts No.	Bolts or Stud Bolts Dia.	F	G
150-lb. Steel										
1/2	1/2	3 1/2	7/16	. . .	1 3/8	2 3/8	4	1/2	2 1/4	1 3/4
3/4	3/4	3 7/8	1/2	. . .	1 11/16	2 3/4	4	1/2	2 1/4	2
1	1	4 1/4	9/16	7/16	2	3 1/8	4	1/2	2 1/2	2
1 1/4	1 1/4	4 5/8	5/8	1/2	2 1/2	3 1/2	4	1/2	2 1/2	2 1/4
1 1/2	1 1/2	5	11/16	9/16	2 7/8	3 7/8	4	1/2	2 3/4	2 1/4
2	2	6	3/4	5/8	3 5/8	4 3/4	4	5/8	3	2 3/4
2 1/2	2 1/2	7	7/8	11/16	4 1/8	5 1/2	4	5/8	3 1/4	3
3	3	7 1/2	15/16	3/4	5	6	4	5/8	3 1/2	3
3 1/2	3 1/2	8 1/2	15/16	13/16	5 1/2	7	8	5/8	3 1/2	3
4	4	9	15/16		6 3/16	7 1/2	8	5/8	3 1/2	3
5	5	10	15/16		7 5/16	8 1/2	8	3/4	3 3/4	3 1/4
6	6	11	1		8 1/2	9 1/2	8	3/4	3 3/4	3 1/4
8	8	13 1/2	1 1/8		10 5/8	11 3/4	8	3/4	4	3 1/2
10	10	16	1 3/16		12 3/4	14 1/4	12	7/8	4 1/2	3 3/4
12	12	19	1 1/4		15	17	12	7/8	4 1/2	4
14	13 1/4	21	1 3/8		16 1/4	18 3/4	12	1	5	4 1/4
16	15 1/4	23 1/2	1 1/16		18 1/2	21 1/4	16	1	5 1/4	4 1/2
18	17 1/4	25	1 9/16		21	22 3/4	16	1 1/8	5 3/4	4 3/4
20	19 1/4	27 1/2	1 11/16		23	25	20	1 1/8	6	5 1/4
24	23 1/4	32	1 7/8		27 1/4	29 1/2	20	1 1/4	6 3/4	5 3/4

Table 3-5. Templates for 150- and 300-lb. Steel Flanges (Cont'd)

Size	A Valve or fitting	B	C Companion flange	C Valve or fitting	D	E	Bolts or Stud Bolts No.	Bolts or Stud Bolts Dia.	F	G
300-lb. Steel										
1/2	1/2	3 3/4	9/16		1 3/8	2 5/8	4	1/2	2 1/2	2
3/4	3/4	4 5/8	5/8		1 11/16	3 1/4	4	5/8	2 3/4	2 1/2
1	1	4 7/8	11/16		2	3 1/2	4	5/8	3	2 1/2
1 1/4	1 1/4	5 1/4	3/4		2 1/2	3 7/8	4	5/8	3	2 3/4
1 1/2	1 1/2	6 1/8	13/16		2 7/8	4 1/2	4	5/8	3 1/2	3
2	2	6 1/2	7/8		3 5/8	5	8	3/4	3 1/4	3
2 1/2	2 1/2	7 1/2	1		4 1/8	5 7/8	8	3/4	3 3/4	3 1/4
3	3	8 1/4	1 1/8		5	6 5/8	8	3/4	4	3 1/2
3 1/2	3 1/2	9	1 3/16		5 1/2	7 1/4	8	3/4	4 1/4	3 3/4
4	4	10	1 1/4		6 3/16	7 7/8	8	3/4	4 1/4	3 3/4
5	5	11	1 3/8		7 5/16	9 1/4	8	3/4	4 1/2	4
6	6	12 1/2	1 7/16		8 1/2	10 5/8	12	3/4	4 3/4	4 1/4
8	8	15	1 5/8		10 5/8	13	12	7/8	5 1/4	4 3/4
10	10	17 1/2	1 7/8		12 3/4	15 1/4	16	1	6	5 1/4
12	12	20 1/2	2		15	17 3/4	16	1 1/8	6 1/2	5 3/4
14	13 1/4	23	2 1/8		16 1/4	20 1/4	20	1 1/8	6 3/4	6
16	15 1/4	25 1/2	2 1/4		18 1/2	22 1/2	20	1 1/4	7 1/4	6 1/2
18	17	28	2 3/8		21	24 3/4	24	1 1/4	7 1/2	6 3/4
20	19	30 1/2	2 1/2		23	27	24	1 1/4	8	7
24	23	36	2 3/4		27 1/4	32	24	1 1/2	9	7 3/4

Table 3-6. Templates for 250- and 800-lb. Hydraulic Cast-Iron Flanges

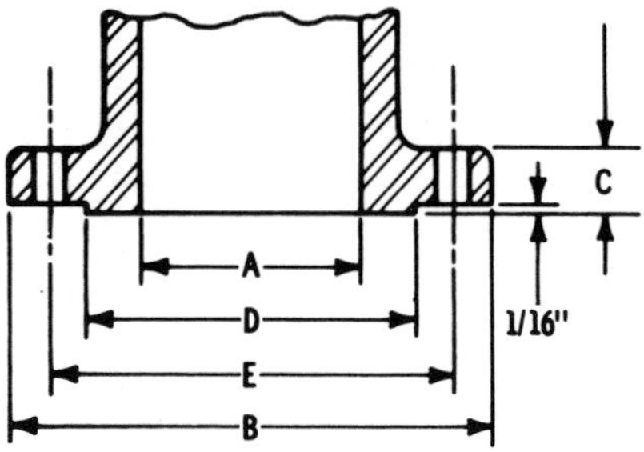

Size	A Valve or fitting	B	C	D	E	No. of bolts	Dia. of bolts	Length of bolts See Note X	Length of stud bolts with 2 nuts See Note X
250-lb. Cast Iron									
1	1	4 7/8	11/16	2 11/16	3 1/2	4	5/8	2 1/2	...
1 1/4	1 1/4	5 1/4	3/4	3 1/16	3 7/8	4	5/8	2 1/2	...
1 1/2	1 1/2	6 1/8	13/16	3 9/16	4 1/2	4	3/4	2 3/4	...
2	2	6 1/2	7/8	4 3/16	5	8	5/8	2 3/4	...
2 1/2	2 1/2	7 1/2	1	4 15/16	5 7/8	8	3/4	3 1/4	...
3	3	8 1/4	1 1/8	5 11/16	6 5/8	8	3/4	3 1/2	...
3 1/2	3 1/2	9	1 3/16	6 5/16	7 1/4	8	3/4	3 1/2	...
4	4	10	1 1/4	6 15/16	7 7/8	8	3/4	3 3/4	...
5	5	11	1 3/8	8 5/16	9 1/4	8	3/4	4	...
6	6	12 1/2	1 7/16	9 11/16	10 5/8	12	3/4	4	...
8	8	15	1 5/8	11 15/16	13	12	7/8	4 1/2	...
10	10	17 1/2	1 7/8	14 1/16	15 1/4	16	1	5 1/4	...
12	12	20 1/2	2	16 7/16	17 3/4	16	1 1/8	5 1/2	...
14	13 1/4	23	2 1/8	18 15/16	20 1/4	20	1 1/8	6	...
16	15 1/4	25 1/2	2 1/4	21 1/16	22 1/2	20	1 1/4	6 1/4	...
18	17	28	2 3/8	23 5/16	24 3/4	24	1 1/4	6 1/2	...
20	19	30 1/2	2 1/2	25 9/16	27	24	1 1/4	6 3/4	...
24	23	36	2 3/4	30 1/4	32	24	1 1/2	7 3/4	9 1/2
30	29	43	3	37 3/16	39 1/4	28	1 3/4	8 1/2	10 1/2
36	34 1/2	50	3 3/8	43 11/16	46	32	2	9 1/2	11 3/4
42	40 1/4	57	3 11/16	50 7/16	52 3/4	36	2	10 1/4	12 1/2
48	46	65	4	58 7/16	60 3/4	40	2	10 3/4	13

Table 3-6. Templates for 250- and 800-lb. Hydraulic Cast-Iron Flanges (Cont'd)

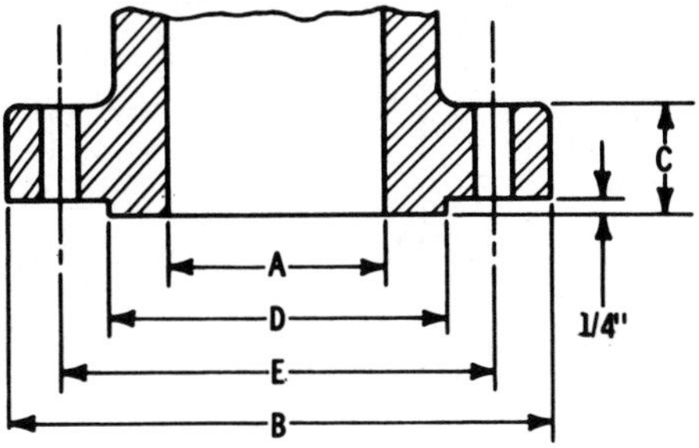

Size	A Valve or fitting	B	C	D	E	No. of bolts	Dia. of bolts	Length of bolts See Note Y	 See Note Z
800-lb. Hydraulic Cast Iron									
2	2	6 1/2	1 1/4	3 5/8	5	8	5/8	3 3/4	3 1/2
2 1/2	2 1/2	7 1/2	1 3/8	4 1/8	5 7/8	8	3/4	4 1/4	4
3	3	8 1/4	1 1/2	5	6 5/8	8	3/4	4 1/2	4 1/4
3 1/2	3 1/2	9	1 5/8	5 1/2	7 1/4	8	7/8	5	4 3/4
4	4	10 3/4	1 7/8	6 3/16	8 1/2	8	7/8	5 1/2	5 1/4
5	5	13	2 1/8	7 5/16	10 1/2	8	1	6	5 3/4
6	6	14	2 1/4	8 1/2	11 1/2	12	1	6 1/4	6
8	7 7/8	16 1/2	2 1/2	10 5/8	13 3/4	12	1 1/8	7	6 3/4
10	9 3/4	20	2 7/8	12 3/4	17	16	1 1/4	7 3/4	7 1/2
12	11 3/4	22	3	15	19 1/4	20	1 1/4	8	7 3/4

Note X—1/16" raised face to 1/16" raised face valve, fitting, or companion flange.
Note Y—1/4" male to 1/4" male valve, fitting, or companion flange.
Note Z—1/4" large male to 3/16" large female valve, fitting, or companion flange.
1/4" tongue to 3/16" groove valve, fitting, or companion flange.

Table 3-7. Templates for 1500- and 2500-lb. Steel Flanges

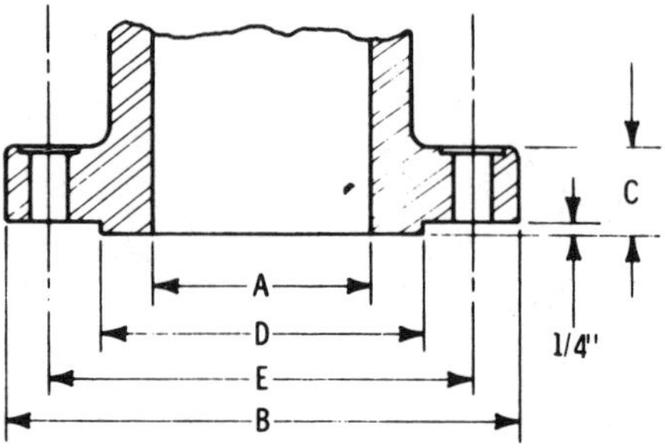

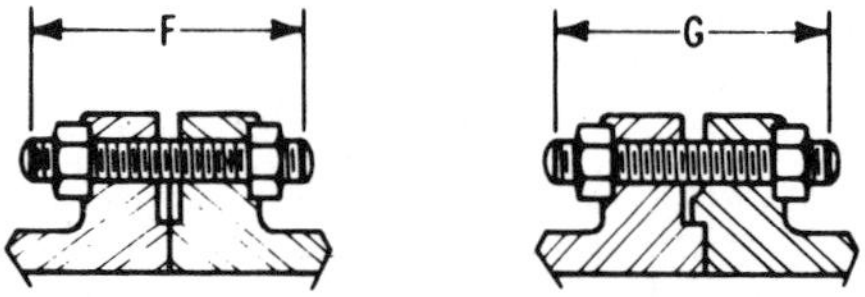

Size	A†	B	C	D	E	Stud Bolts		F	G
						No.	Dia.		
1500-lb. Steel									
1/2	1/2	4 3/4	7/8	1 3/8	3 1/4	4	3/4	4	3 3/4
3/4	11/16	5 1/8	1	1 11/16	3 1/2	4	3/4	4 1/4	4
1	7/8	5 7/8	1 1/8	2	4	4	7/8	4 3/4	4 1/2
1 1/4	1 1/8	6 1/4	1 1/8	2 1/2	4 3/8	4	7/8	4 3/4	4 1/2
1 1/2	1 3/8	7	1 1/4	2 7/8	4 7/8	4	1	5 1/4	5
2	1 7/8	8 1/2	1 1/2	3 5/8	6 1/2	8	7/8	5 1/2	5 1/4
2 1/2	2 1/4	9 5/8	1 5/8	4 1/8	7 1/2	8	1	6	5 3/4
3	2 3/4	10 1/2	1 7/8	5	8	8	1 1/8	6 3/4	6 1/2
4	3 5/8	12 1/4	2 1/8	6 3/16	9 1/2	8	1 1/4	7 1/2	7 1/4
5	4 3/8	14 3/4	2 7/8	7 5/16	11 1/2	8	1 1/2	9 1/2	9 1/4
6	5 3/8	15 1/2	3 1/4	8 1/2	12 1/2	12	1 3/8	10	9 3/4
8	7	19	3 5/8	10 5/8	15 1/2	12	1 5/8	11 1/4	11
10	8 3/4	23	4 1/4	12 3/4	19	12	1 7/8	13 1/4	13
12	10 3/8	26 1/2	4 7/8	15	22 1/2	16	2	14 3/4	14 1/2
14	11 3/8	29 1/2	5 1/4	16 1/4	25	16	2 1/4	16	15 3/4
16	13	32 1/2	5 3/4	18 1/2	27 3/4	16	2 1/2	17 1/2	17 1/4
18	14 5/8	36	6 3/8	21	30 1/2	16	2 3/4	19 1/4	19
20	16 3/8	38 3/4	7	23	32 3/4	16	3	21	20 3/4
24	19 5/8	46	8	27 1/4	39	16	3 1/2	24	23 3/4

† Dimension "A" applies to valves or fittings.

Table 3-7.
Templates for 1500- and 2500-lb. Steel Flanges (Cont'd)

Size	A†	B	C	D	E	Stud Bolts No.	Stud Bolts Dia.	F	G
2500-Lb. Steel									
1/2	7/16	5 1/4	1 3/16	1 3/8	3 1/2	4	3/4	4 3/4	4 1/2
3/4	9/16	5 1/2	1 1/4	1 11/16	3 3/4	4	3/4	4 3/4	4 1/2
1	3/4	6 1/4	1 3/8	2	4 1/4	4	7/8	5 1/4	5
1 1/4	1	7 1/4	1 1/2	2 1/2	5 1/8	4	1	5 3/4	5 1/2
1 1/2	1 1/8	8	1 3/4	2 7/8	5 3/4	4	1 1/8	6 1/2	6 1/4
2	1 1/2	9 1/4	2	3 5/8	6 3/4	8	1	6 3/4	6 1/2
2 1/2	1 7/8	10 1/2	2 1/4	4 1/8	7 3/4	8	1 1/8	7 1/2	7 1/4
3	2 1/4	12	2 5/8	5	9	8	1 1/4	8 1/2	8 1/4
4	2 7/8	14	3	6 3/16	10 3/4	8	1 1/2	9 3/4	9 1/2
5	3 5/8	16 1/2	3 5/8	7 5/16	12 3/4	8	1 3/4	11 1/2	11 1/4
6	4 3/8	19	4 1/4	8 1/2	14 1/2	8	2	13 1/2	13 1/4
8	5 3/4	21 3/4	5	10 5/8	17 1/4	12	2	15	14 3/4
10	7 1/4	26 1/2	6 1/2	12 3/4	21 1/4	12	2 1/2	19	18 3/4
12	8 5/8	30	7 1/4	15	24 3/8	12	2 3/4	21	20 3/4

†Dimension "A" applies to valves or fittings.

Union Elbow and Tees

The frequent use of unions in pipe lines is desirable for convenience in case of repairs. Where the union is combined with a fitting, the advantage of a union is obtained with only one threaded joint instead of two, as in the case of a separate union. A disadvantage of union fittings is that they are not usually as easily obtainable as ordinary fittings. Fig. 3-18 shows various union elbows and union tees of the female and of the male and female types.

Expansion Joints

The linear expansion and contraction of pipe, due to differences in the temperature of the fluid carried and of the surrounding air, must be cared for by suitable expansion joints or bends. Table 3-8 is provided in order to determine the amount of expansion or contraction in a pipe line, showing the increase in length of a pipe 100 ft. long at various temperatures. The expansion of any length of pipe may be found by taking the difference in increased length

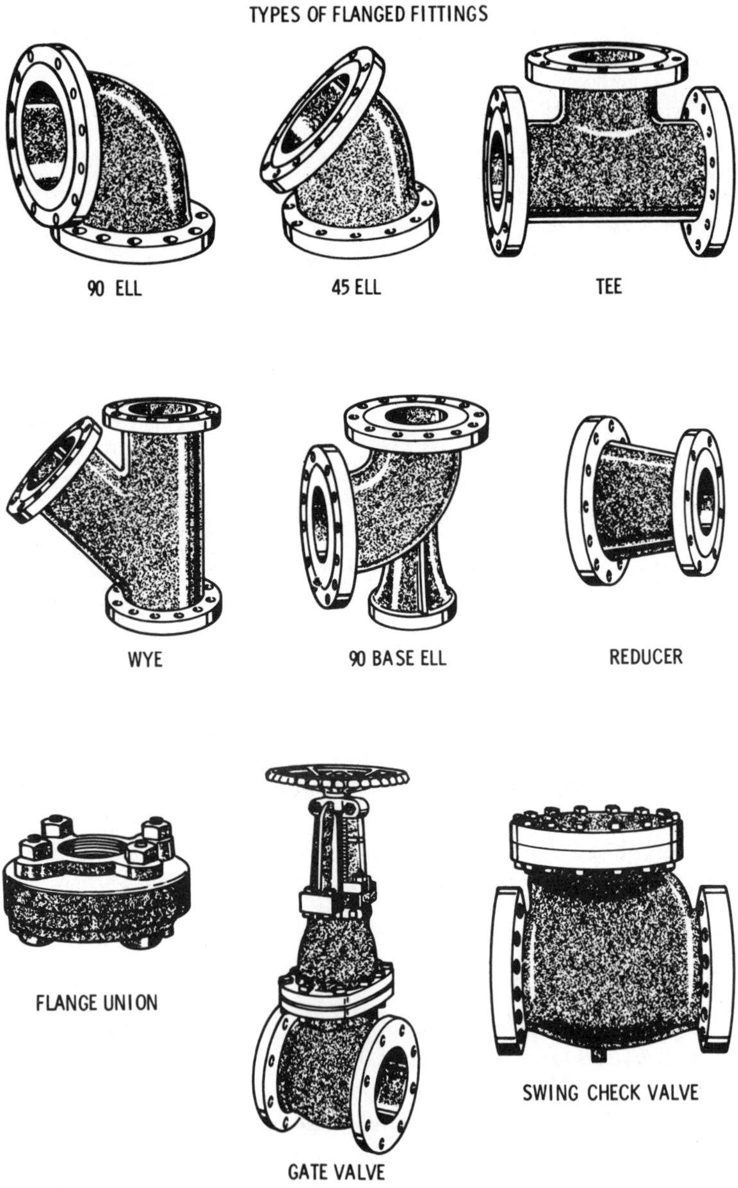

Fig. 3-17. Industrial flanges.

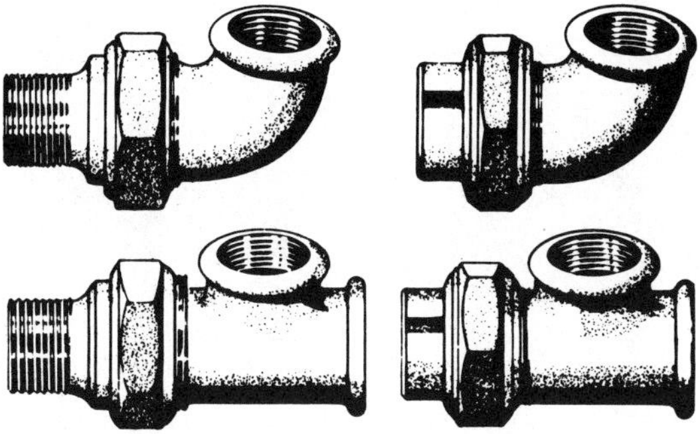

Fig. 3-18. Various union elbows and union tees.

at the minimum and maximum temperatures, dividing by 100 and multiplying by the length of the line under consideration.

Drainage Fittings

Drainage fittings of cast iron, copper, and plastic differ slightly from ordinary fittings. They are so constructed that when the pipe is made into the fitting, there is a smooth unobstructed passageway for liquid and water-carried wastes. If standard water fittings were used for drainage, there would be a shoulder at the point where the pipe stops in the fitting, and this shoulder would obstruct the flow through the pipe. Copper and plastic fittings are better for drainage purposes because they extend completely into the fitting recess; when one uses threaded fittings, the pipe may not screw into the recess completely when tight and may leave a shoulder or ridge. Drainage fittings are shown and compared with a standard fitting in Fig. 3-19.

Soil Pipe and Fittings

Hub-type soil pipe using rubber gaskets (Fig. 3-20) instead of lead joints between pipe fittings is widely used. The rubber gaskets are color coded to match the weight of soil pipe being used. A special tool, shown in Fig. 3-21, can be used to pull the pipe and fittings together. Soil pipe and fittings that have a calking bead on the spigot end cannot be used with the rubber joint.

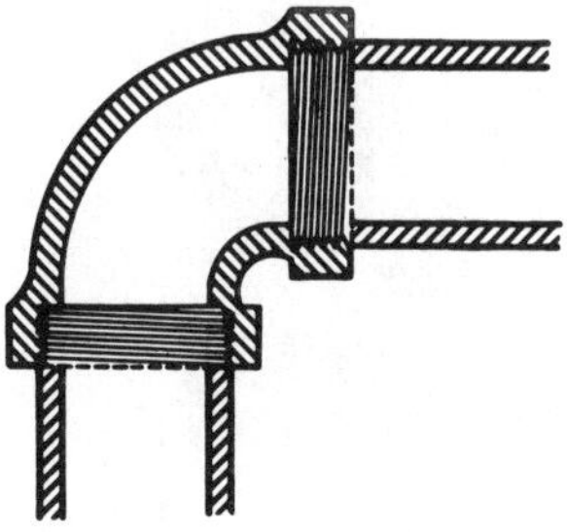

(A) Threaded drainage fitting—smooth passageway.

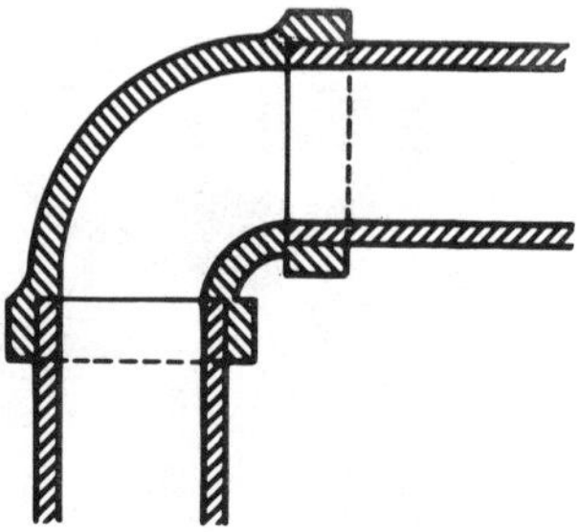

(B) Copper or plastic drainage fitting smooth passageway.

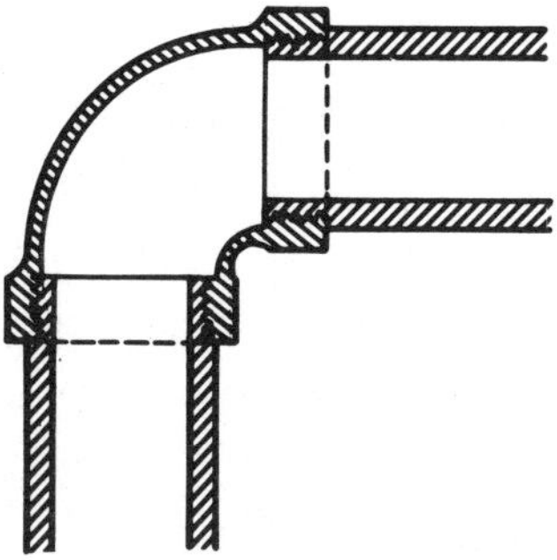

(C) Standard pipe fitting—shoulder on pipe at makeup point

Fig. 3-19. Comparison of standard and drainage fittings.

Fig. 3-20. Type of gasket used with compression-type soil pipe.

Table 3-8. Expansion of Steam Pipes
(Inches Increase per 100 feet)

Temperature (Degrees F.)	Steel	Wrought Iron	Cast Iron	Brass and Copper
0	0.00	0.00	0.00	0.00
20	0.15	0.15	0.10	0.25
40	0.30	0.30	0.25	0.45
60	0.45	0.45	0.40	0.65
80	0.60	0.60	0.55	0.90
100	0.75	0.80	0.70	1.15
120	0.90	0.95	0.85	1.40
140	1.10	1.15	1.00	1.65
160	1.25	1.35	1.15	1.90
180	1.45	1.50	1.30	2.15
200	1.60	1.65	1.50	2.40
220	1.80	1.85	1.65	2.65
240	2.00	2.05	1.80	2.90
260	2.15	2.20	1.95	3.15
280	2.35	2.40	2.15	3.45
300	2.50	2.60	2.35	3.75
320	2.70	2.80	2.50	4.05
340	2.90	3.05	2.70	4.35
360	3.05	3.25	2.90	4.65
380	3.25	3.45	3.10	4.95
400	3.45	3.65	3.30	5.25
420	3.70	3.90	3.50	5.60
440	3.95	4.20	3.75	5.95
460	4.20	4.45	4.00	6.30
480	4.45	4.70	4.25	6.65
500	4.70	4.90	4.45	7.05
520	4.95	5.15	4.70	7.45
540	5.20	5.40	4.95	7.85
560	5.45	5.70	5.20	8.25
580	5.70	6.00	5.45	8.65
600	6.00	6.25	5.70	9.05
620	6.30	6.55	5.95	9.50
640	6.55	6.85	6.25	9.95
660	6.90	7.20	6.55	10.40
680	7.20	7.50	6.85	10.95
700	7.50	7.85	7.15	11.40
720	7.80	8.20	7.45	11.90
740	8.20	8.55	7.80	12.40
760	8.55	8.90	8.15	12.95
780	8.95	9.30	8.50	13.50
800	9.30	9.75	8.90	14.10

Courtesy Ridge Tool Co.

Fig. 3-21. Soil-pipe assembly tool used to install compression soil pipe and fittings.

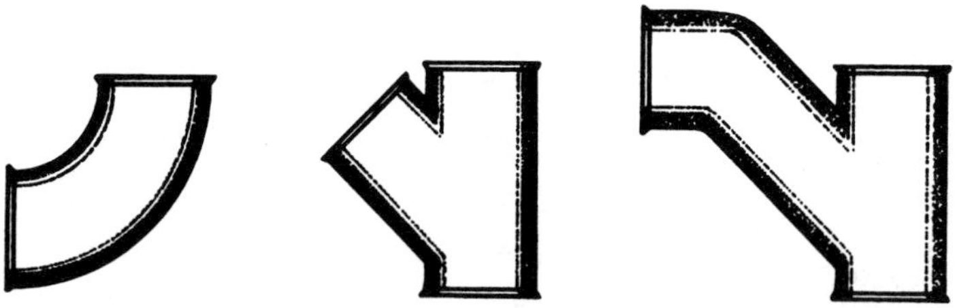

(A) 1/4 bend. (B) Wye. (C) Combination wye and 1/8 bend.

Fig. 3-22. NO-HUB® soil-pipe fittings: (A) 1/4 bend, (B) wye, and (C) combination wye and 1/8 bend.

NO-HUB® soil pipe fittings are made in standard patterns as shown in Fig. 3-22. Connections are made between pipe and fittings using stainless steel clamps with neoprene gaskets as shown in Fig. 3-23. A special torque wrench, such as the one shown in Fig. 3-24, is available for tightening clamps on NO-HUB® soil pipe.

There are a number of typical fittings for use with lead and oakum soil-pipe installations. These are shown in Fig. 3-25.

Courtesy Cast-Iron Soil Pipe Institute

Fig. 3-23. A typical connection made between pipe and fittings using stainless steel clamps with neoprene gaskets.

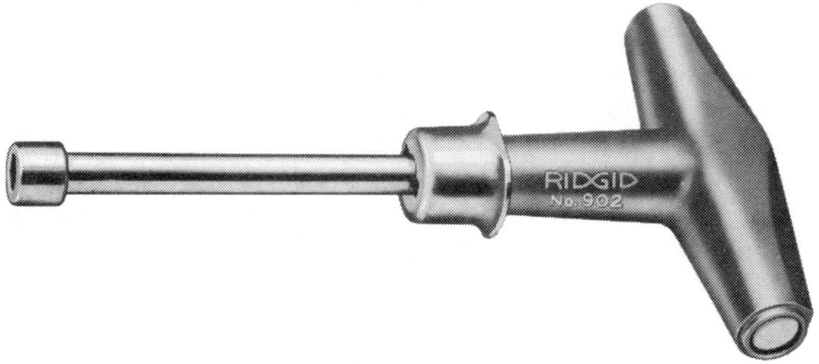

Courtesy Ridge Tool Co

Fig. 3-24. A torque wrench used for tightening NO-HUB® soil-pipe.

Special Fittings for Wall-hung Toilets

Special fittings may be obtained for use wherever back- or wall-outlet toilets are installed in batteries. They are especially adapted for use in buildings of reinforced-concrete construction. Using these fittings in connection with wall-hung toilets eliminates the necessity of cutting and thus weakening the floors, as the horizontal waste line is entirely above the floor. Before the advent of these fittings, it was always necessary to suspend the horizontal waste line of a battery of toilets from the ceiling below, unless a groove was made in the floor or the floor of the toilet room raised. All these methods are objectionable but necessary where ordinary drainage fittings are used. As will be seen from the accompanying illustrations, these fittings are tapped for the toilet connection at different distances from the center of the run, so that when the toilets in a battery are set in line and the fittings placed in consecutive order according to the tapping numbers given them, the waste line is given a pitch.

The fittings shown in Figs. 3-26 and 3-27 are made in R.H. (right hand) and L.H. (left hand) flow patterns. To identify the fitting correctly, place the fitting as shown in Fig. 3-26, facing the fixture inlet. If the spigot end is on the right, the fitting is a R.H. flow

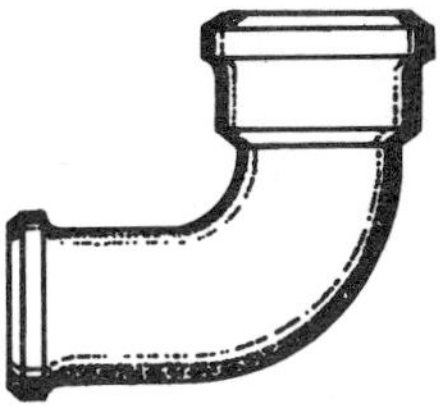

(A) Short pattern $\frac{1}{4}$ bend.

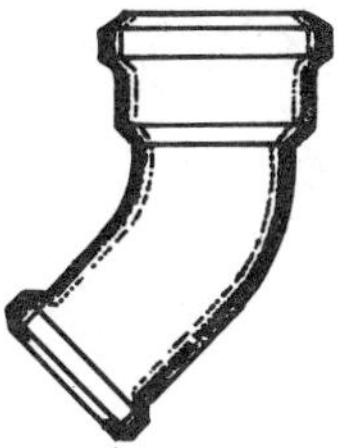

(B) Short pattern $\frac{1}{8}$ bend.

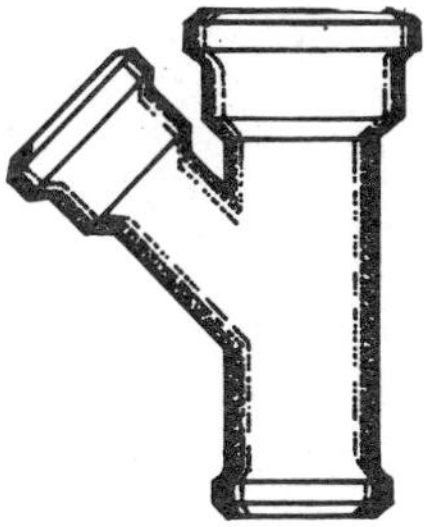

(C) Wye (Y).

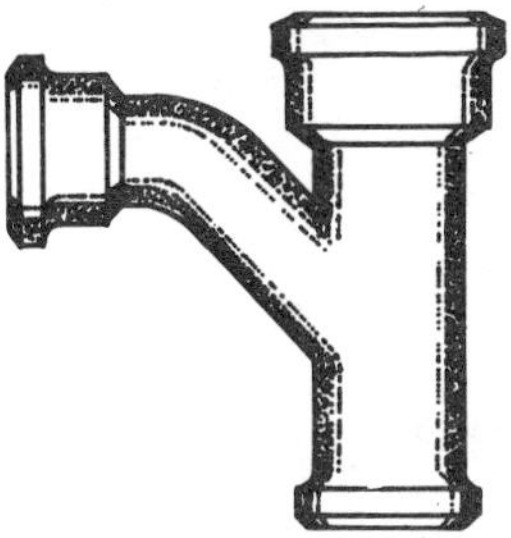

(D) Combination wye and $\frac{1}{8}$ bend.

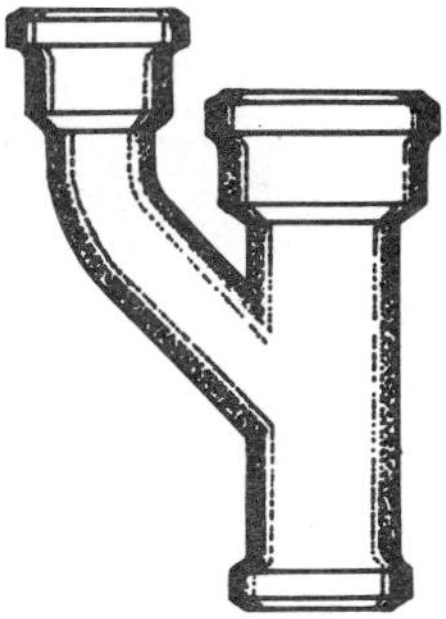

(E) Single upright wye branch.

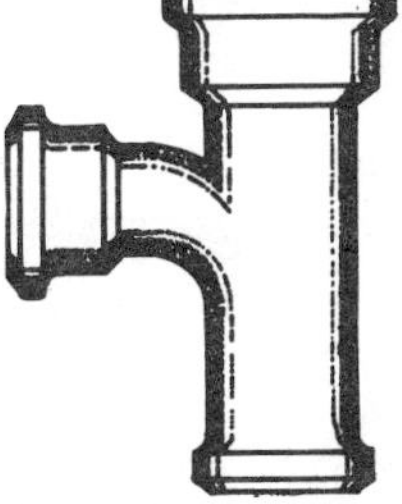

(F) Sanitary tee.

(G) Roof increaser.

Fig. 3-25. Typical fittings used with lead and oakum soil-pipe installation.

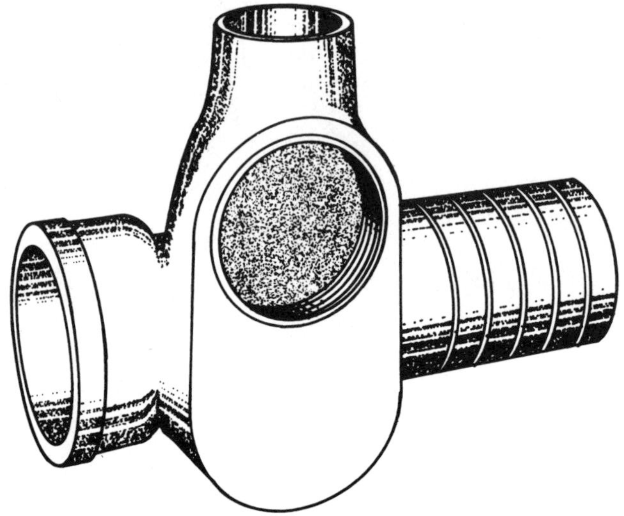

Fig. 3-26. A 4-in. tapping shown in position 1.

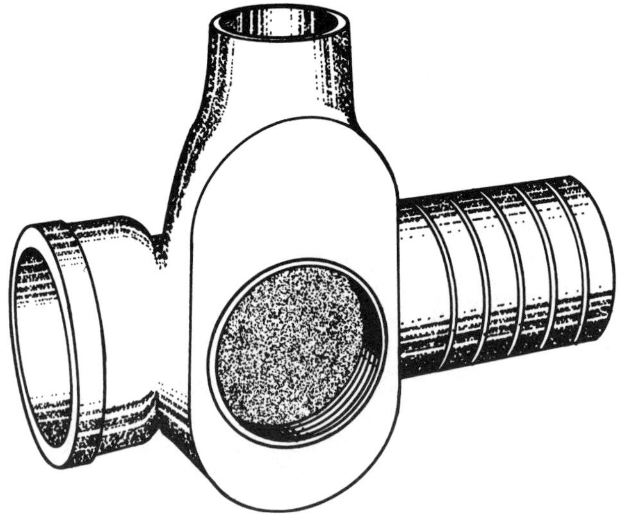

Fig. 3-27. A 4-in. tapping shown in position 5.

fitting; if the spigot is on the left, the fitting is a L.H. flow fitting. As pictured in Fig. 3-26, the fitting is a R.H. flow fitting.

The tables for figuring the tapping numbers for the special fittings shown in Fig. 3-26 are found in the Tapping and Dimension tables at the end of Chapter 6, Roughing-in.

CHAPTER 4

Pipe Joints

Before taking up roughing-in work, the plumbing student should know something about the different kinds of pipe used in the trade and the large variety of fittings used. He should have a working knowledge of the joints used with various types of pipe and fittings, how the joints are made, and the reason for using certain fittings or materials.

SOIL PIPE

There are several types of soil pipe used for drainage and vent piping, depending on the particular job requirements, the specifications written by the architect and engineer, and the plumbing code governing the location of the job. Three types of joints may be used:

1. The lead and oakum joint.
2. The compression joint.
3. The NO-HUB® joint.

Cast-iron soil-pipe joints, when made properly, are water and gas tight, semirigid connections of two or more pieces of pipe or fittings in a sanitary system.

LEAD AND OAKUM JOINTS

One way of assembling cast-iron soil pipe is to use lead and oakum joints. Fig. 4-1 shows a spigot end of cast-iron soil pipe inserted into a soil pipe hub, ready for the joint to be made. The joint is made by first packing the oakum into the space between the pipe and the hub (Fig. 4-2), and then pouring molten lead on tip of the oakum, up to the top of the hub. The joint is then calked, or tightened, using a hammer and inside and outside calking irons.

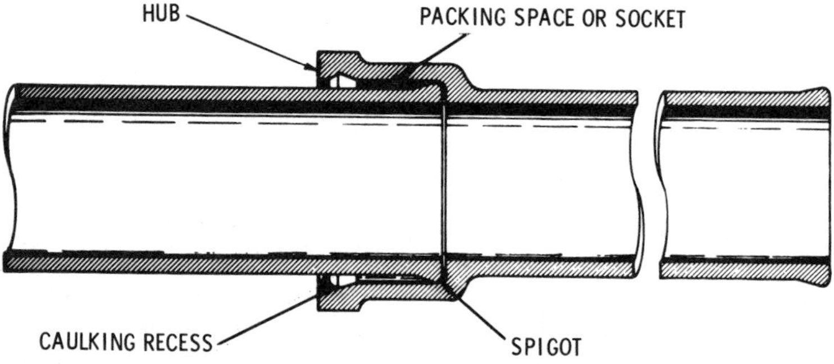

Fig. 4-1. Hub-and-spigot joint used for connecting lengths of cast-iron soil pipe.

There are two types of oakum now on the market, one made of hemp fibers, which are oiled and twisted, and the other a white oakum made of fiber and impregnated with a powdery substance that swells when brought into contact with water. White oakum is preferred by most plumbers because the joint made with it is virtually leakproof. Whichever type of oakum is used, the joint should be tightly packed, using a hammer and yarning or packing iron (Fig. 4-3), and filling the hub to within one inch of the top with oakum. One inch of lead is the correct amount of lead for a joint. An ordinary, accessible joint will only require the use of a regular yarning iron, if the joint is in a corner, special yarning and calking irons may be needed, as shown in Fig. 4-4. Numerous other cases of close work occur, where various-shaped tools are made to meet all conditions. Some of the special tools generally used are shown in Fig. 4-5.

For joints near the ceiling, it is necessary to use a ceiling-drop tool to pack the socket, as shown in Fig. 4-6. The handle of this tool is quite heavy, so that the yarn may be forced into the socket by a series of light blows with the hand, as in Fig. 4-6A. The offset at the handle provides a surface for blows with a hammer in packing the yarn tightly in the socket, as shown in Fig. 4-6B. Table 4-1 shows the amount of oakum and lead needed for making up joints on various sizes of soil pipes.

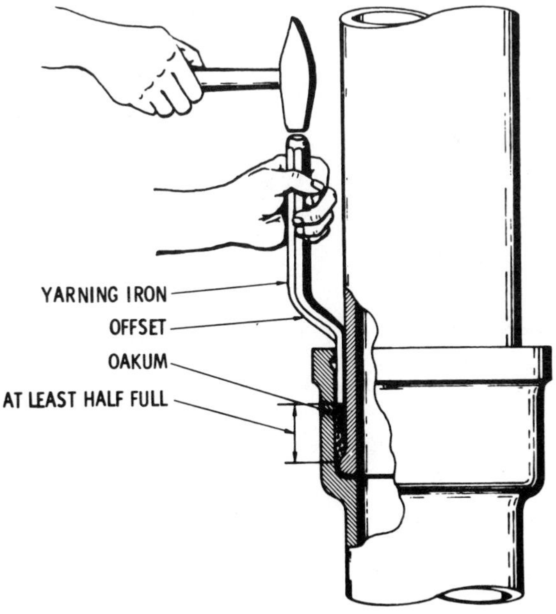

Fig. 4-2. Packing oakum into the hub and spigot of a sewer pipe.

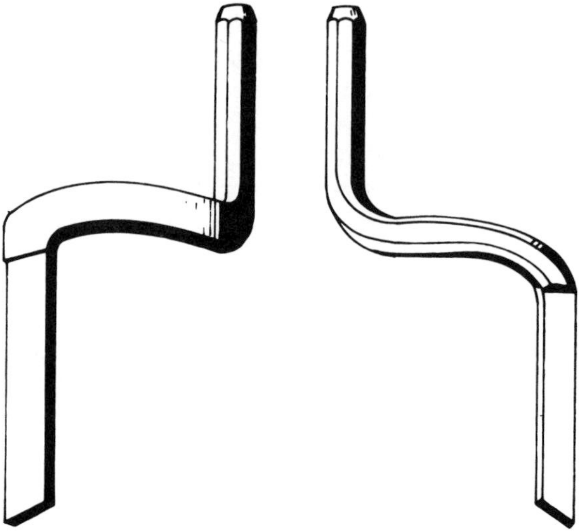

Fig. 4-3. Left-hand (left) and right-hand (right) yarning irons.

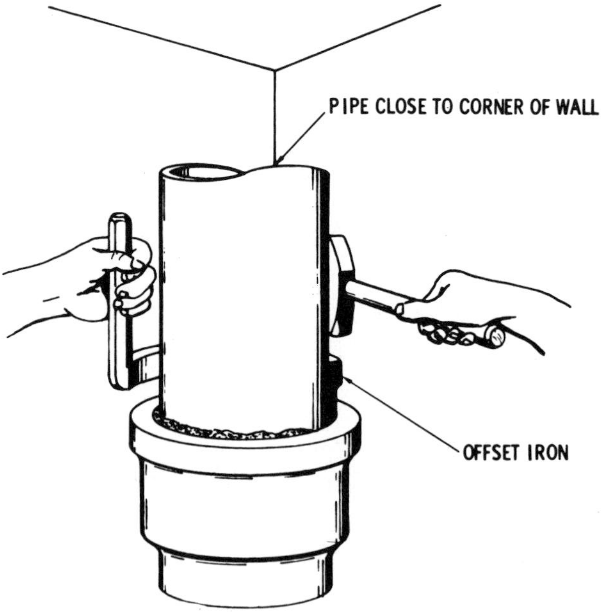

Fig. 4-4. Using a special yarning iron to pack the oakum where the pipe is close to a wall.

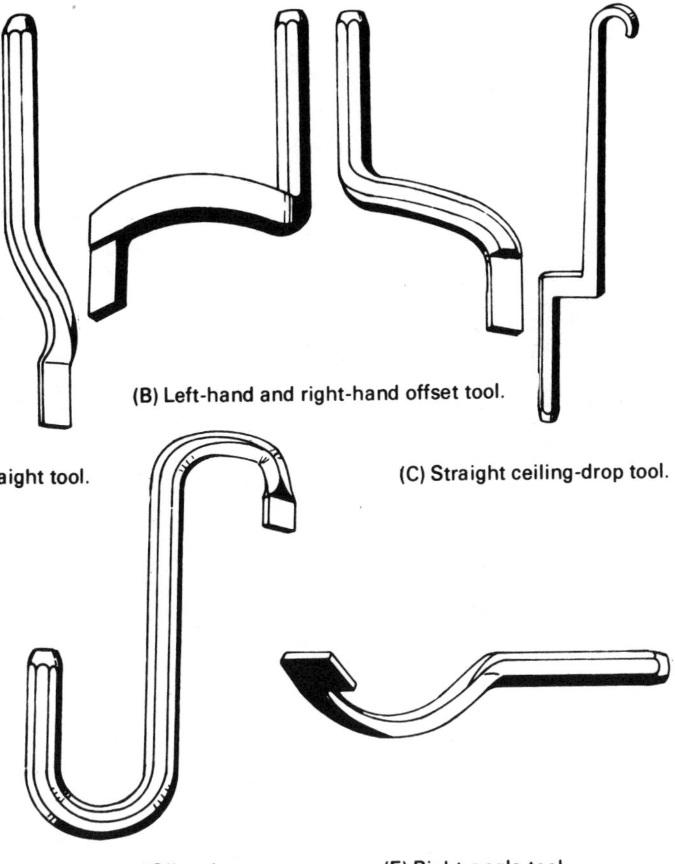

(A) Straight tool.

(B) Left-hand and right-hand offset tool.

(C) Straight ceiling-drop tool.

(D) Ceiling-drop "S" tool.

(E) Right-angle tool.

Fig. 4-5. Special yarning and calking tools.

Pouring the Lead

When the yarn has been evenly and tightly compressed all around in the joint, the next operation is pouring the lead. Table 4-1 lists the amount of lead required for the various sizes of pipe. The actual amount of lead required will, of course, depend on the proportion of oakum and lead put into the socket. About one lb. of lead is required for each inch of diameter of the pipe for each average joint. It is important to fill the socket at one pouring. First,

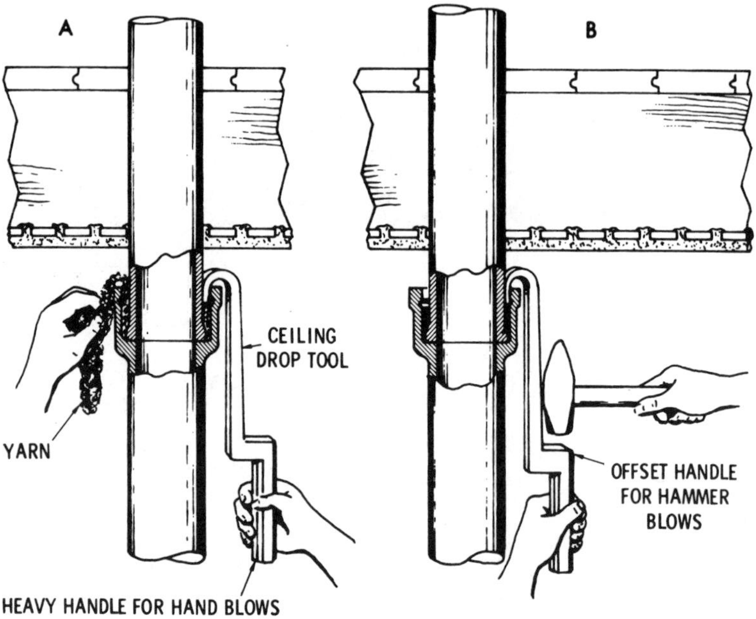

Fig. 4-6. Method of using a ceiling-drop tool.

Table 4-1. Oakum and Lead Requirements for Caulked Joints

Material	Size of Pipe (inches)							
	2	3	4	5	6	7	8	10
Oakum (feet)	3	4½	5	6½	7½	8½	9½	12
Lead (lbs.)	1½	2¼	3	3¾	4½	5¼	6	7½

melt plenty of lead, and then with the pouring ladle dip out a supply ample to fill the socket without a second dipping.

Before pouring, care should be taken to see that there are no projecting strands of oakum, otherwise when the lead is poured, these strands will be consumed, leaving tiny ducts through the lead which could cause leakage. Care should be taken to make sure that the socket is dry before pouring, as the molten lead will turn any moisture into steam, causing an explosion. Such an explosion will hurl the lead out of the joint with considerable force, possibly injuring the plumber. To guard against an injury of this sort, the plumber should stand as far away from the ladle as possible and *out of range* of the direction in which the lead would fly. If is is necessary to pour a wet joint, first pack the oakum tightly, then sprinkle in a teaspoonful of powdered rosin, or oil if rosin is not available. The object of this is to prevent the molten lead from flying when it strikes the moisture. Extreme caution, however, should always be taken when pouring lead.

When lead is to be poured into a horizontal joint, a *joint runner*, as shown in Fig. 4-7, is used. The end of the socket is closed by the joint runner all around except at the top, so that when the molten lead is poured into the socket through this small opening, it will not escape, but will be held in the socket until it cools and solidifies. The runner is then removed and the lead caulked to ensure a tight joint. To avoid possible injury from flying lead, the plumber should stand out of range, the same as when pouring a vertical joint. It is sometimes necessary to pour a joint upside down. This may be done by placing the joint runner around the pipe and clamping as shown in Fig. 4-8, a pouring gate being formed by building up walls of fire clay between the runner and the hub.

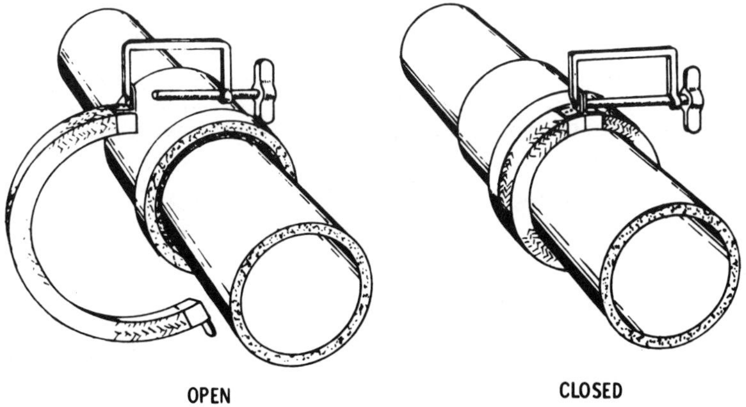

Fig. 4-7. An asbestos joint runner.

Fig. 4-8. Pouring an upside-down joint with the use of a joint runner.

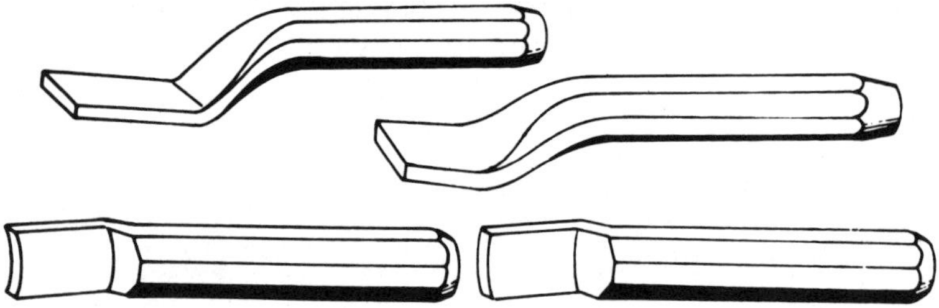

Fig. 4-9. Various calking tools.

Calking the Lead

After pouring the lead, the next operation is calking, which is done with a calking tool. These are similar to yarning tools, except that the blade is shorter and heavier. Some plumbers calk while the joint is hot, others after is has cooled. The best method is to calk moderately tight while the joint is hot so that the lead will better adjust itself to any irregularities of the socket walls. After the joint has cooled, the calking is finished by driving the lead into contact with the spigot surface on one edge and against the inner surface of the hub on the other. Where the joint is fully accessible, regular pattern tools are used, as shown in Fig. 4-9. In addition there are numerous special tools to facilitate calking in close places. Fig. 4-10 shows some of the shapes frequently employed and how they are used.

CUTTING SOIL PIPE

In any job of fitting soil pipe there will be numerous places where it is necessary to cut a length of pipe to make up the line. This is because the pipe is cast in standard lengths (usually 5 ft.), and unless the distance between the first and last joints of a line is a multiple of 5, there will be an odd length of pipe needed to complete the line.

A full 5-ft. length of pipe will have a hub on one end and a spigot on the other, as shown in Fig. 4-11, and is called a *single-hub* pipe. However, if a length of a single pipe is cut to obtain a short length, the spigot end will be of no use, resulting in waste, To avoid this, a *double-hub* pattern pipe, as shown in Fig. 4-12, is used. When this is cut, each piece will have a hub, so that the piece can be used. Accordingly, in ordering pipe for any installation, a few lengths of double-hub pattern pipe should be included to avoid waste in cutting.

In some instances, though it is difficult, a chain pipe cutter may not be available, and it will be necessary to cut soil pipe with hammer and chisel. To do so, first make a chalk mark entirely around the pipe where it is to be cut. This mark should be true, not rambling. The pipe should be firmly supported on the floor with a block at the cutting line, as in Fig. 4-13, or preferably on a mound of earth, as shown in Fig. 4-14. The cutting is done as in Fig. 4-15. When using a chisel and hammer, the chisel should be narrow and sharply pointed and the hammer of medium weight.

A difficulty encountered in making up a joint with a cut piece of double-hub pipe is that there is no spigot or bead on the end to center the pipe. Care must be taken to keep the cut end centered in

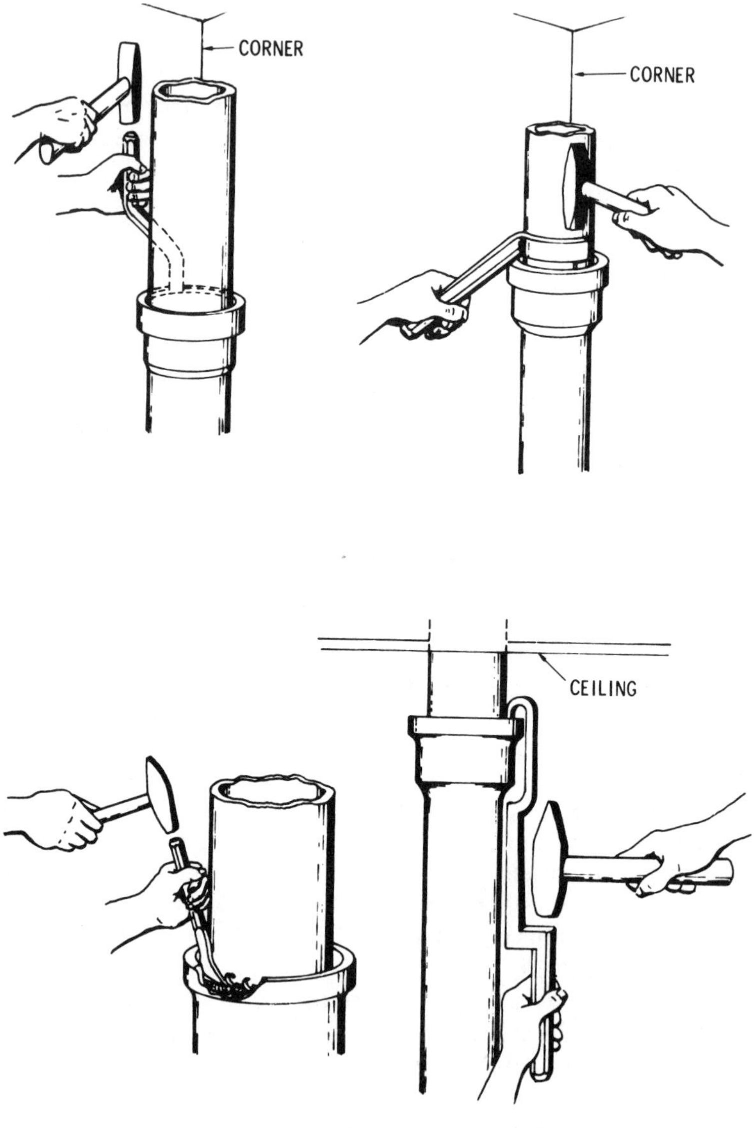

Fig. 4-10. Special calking tools and their uses.

Fig. 4-11. A full length of pipe showing the hub and spigot.

Fig. 4-12. A double-hub pipe.

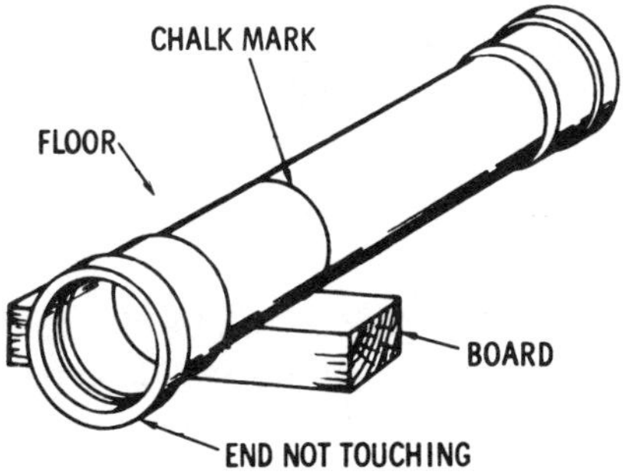

Fig. 4-13. Cutting the double-hub pipe using a piece of wood to support the pipe.

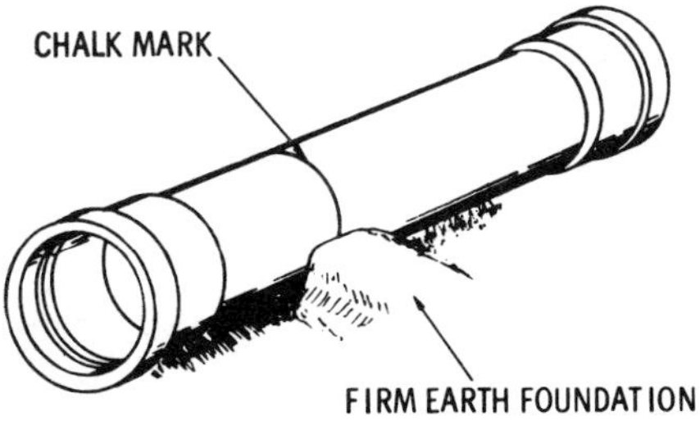

Fig. 4-14. Cutting double-hub pipe, using the ground for support along the cutting line.

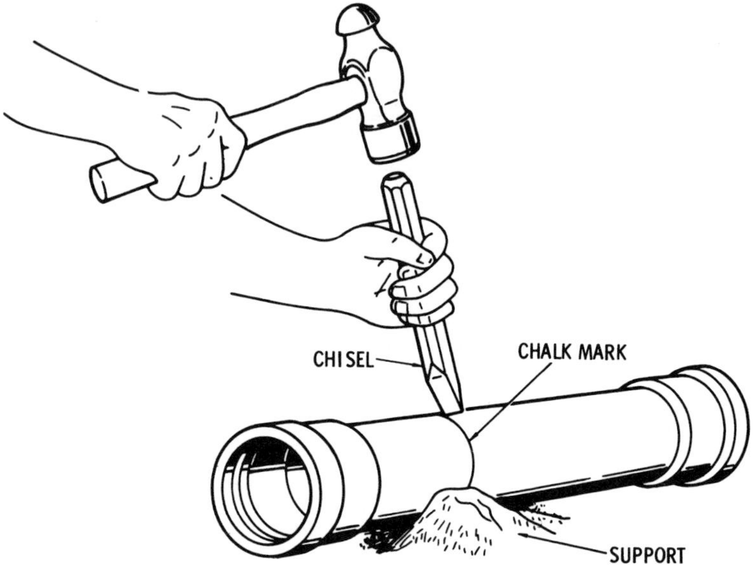

Fig. 4-15. Method of cutting cast-iron soil pipe using a hammer and chisel.

the mating hub so that the packing will be of uniform thickness all around. If the cut end is pushed to one side in packing, it will be difficult to make a tight joint. Fig. 4-16 shows the right and wrong way.

The tool shown in Fig. 4-17 is better and safer for use in cutting soil pipe. There is no danger of cast-iron splinters, as with a hammer and chisel, and the cut will be square and true if the cutter is set properly.

COMPRESSION JOINTS

The compression joint is relatively new in terms of its use with cast-iron soil pipe; similar compression type gaskets have been successfully used with water main for more than thirty years. As shown in Fig. 4-18, the compression joint uses hub-and-spigot pipe, as does the lead and oakum joint. The major differences are the one-piece rubber gasket and the spigot end of the pipe, which

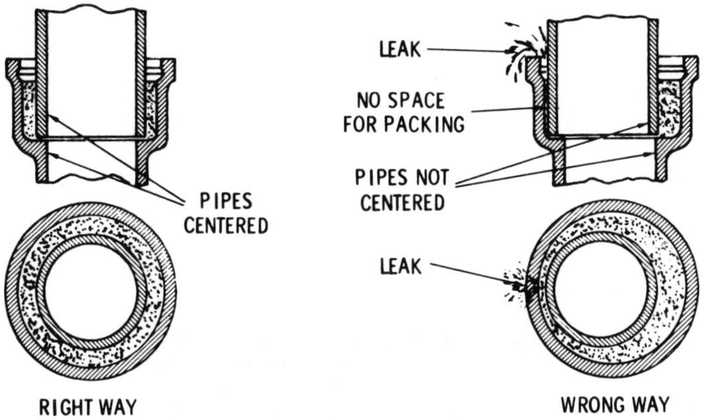

Fig. 4-16. The right way and the wrong way of making a joint.

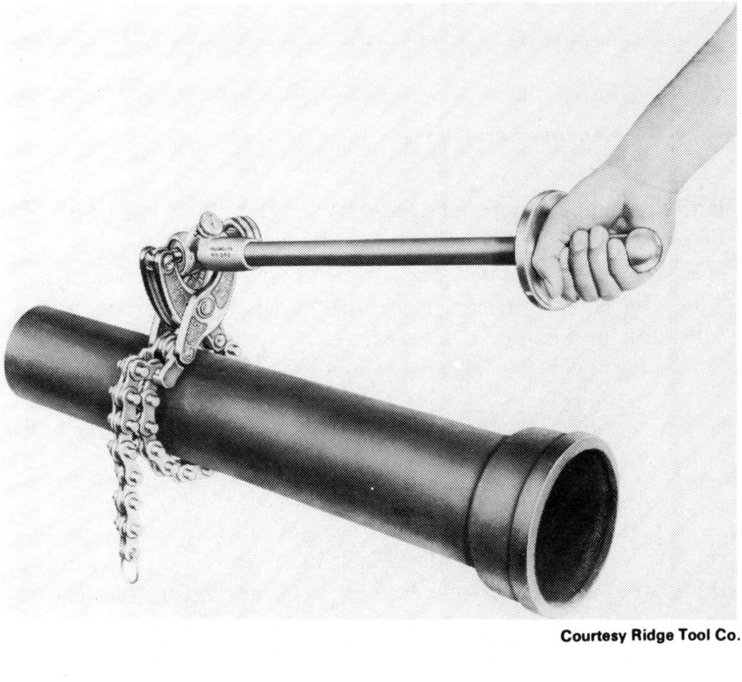

Courtesy Ridge Tool Co.

Fig. 4-17. Ratchet tool and pipe cutter.

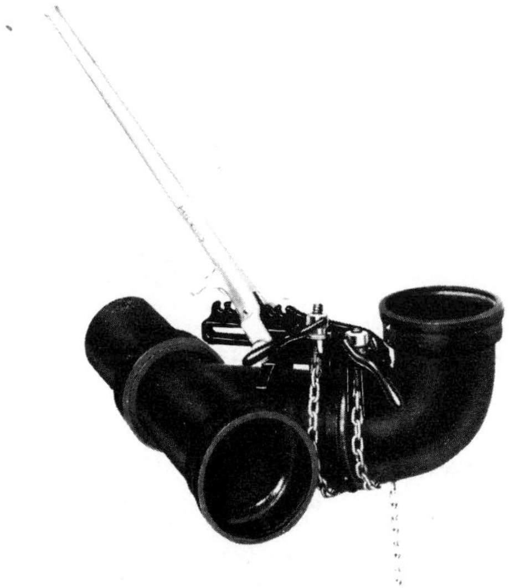

Courtesy Ridge Tool Co.

Fig. 4-18. A compression joint being assembled with special tool.

is always plain or without a bead. When the spigot end of the pipe or fitting is pushed or drawn into the gasketed hub, the joint is sealed by displacement and compression of the rubber gasket. The joint is leak-proof, root-proof, and pressure proof, absorbs vibration, and can be deflected.

THE NO-HUB® JOINT

The NO-HUB® joint for cast-iron soil pipe and fittings is also a relatively new way of joining cast iron. As can be seen in Fig. 4-19, the system uses a one-piece neprene gasket and a stainless-steel shield and retaining clamps. The advantage is that it permits joints to be made against a ceiling or in any limited-access area. In its 2-in. and 3-in. sizes it will fit into a standard 2 × 4 in. partition without furring.

Courtesy Cast-Iron Soil Pipe Institute

Fig. 4-19. A typical connection made between pipe and fittings using stainless steel clamps with neoprene gaskets.

COATED CAST-IRON PIPE

Standard pipe is dipped in hot asphaltum by the manufacturer to prevent the deteriorating effects of corrosion and to fill up any sand holes, flaws, or other defects that may have occured in manufacture.

CORROSIVE-WASTE PIPE

The rapid growth of the use of acids and other corrosives in industrial work, as well as the increasing number of schools, colleges, and hospitals containing chemical laboratories, make it necessary to have a knowledge of special plumbing materials. Among industrial users, the most common are photoengravers, manufacturing jewelers, and industries that manufacture enameled or plated articles and therefore must use acids for cleaning the material.

Several kinds of pipe are used to meet the severe requirements of drainage systems for corrosive wastes, such as:

1. Noncorrosive metal.
2. Plastic.
3. Glass.

Noncorrosive Pipe

An example of noncorrosive pipe is Duriron pipe. This pipe may be cut with a cold chisel and hammer, just like cast-iron soil pipe, but the metal is so hard that the chisel will make only a slight scratch on the pipe. This makes it necessary to go around the pipe two or three times with the chisel and to use about the same weight hammer blows as with cast iron before the pipe will break clean. A pipe cutter having a coil spring above the specially hardened cutter wheel will save much time on a job.

In making joints on Duriron pipe, asbestos rope (at least 85 percent pure) should be used in place of hemp or oakum in order to make an acid-proof joint, and the lead should be poured at as low a temperature as possible. If too hot, the bell or hub may be cracked while calking.

Plastic

Plastic pipe is heavily used particularly in locations where highly-corrosive waste must be disposed of. Industrial plants are using plastic pipe for their water-distribution systems, particularly for underground lawn-sprinkling piping. Homes are equipped with plastic waste pipe and accessories, including P-traps. Plastic is also used for water in residential systems. Plastic pipe is obtainable in *rigid, semirigid,* and *flexible* types and in sizes ranging from ½ to 6 in. in diameter.

Glass

Corrosion and leakage are the two worst problems encountered in chemical-waste drain lines. The use of borosilicate type of glass tubing for drain lines is becoming standard practice where excessive chemical corrosion may occur. Positive lead-free seals are easily made, and no lubricant or sealant is needed in the installation procedures.

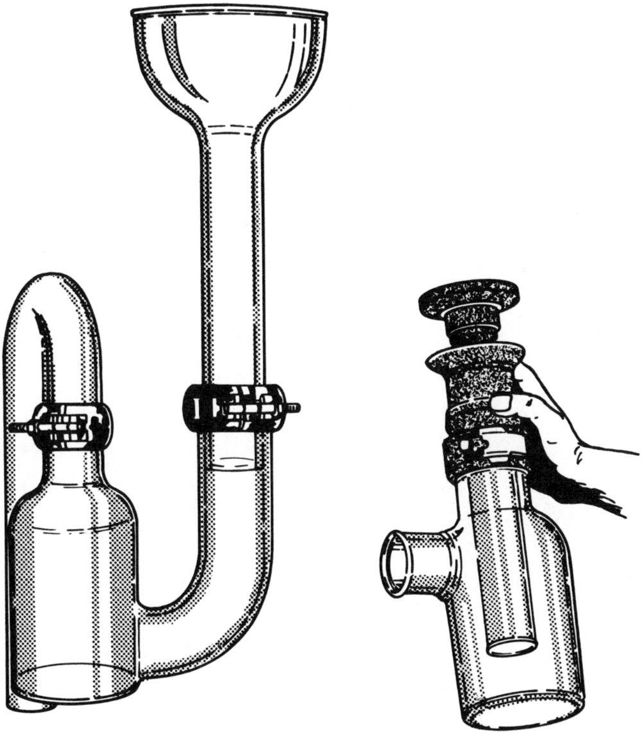

Fig. 4-20. A typical "S"-type drum trap made of glass.

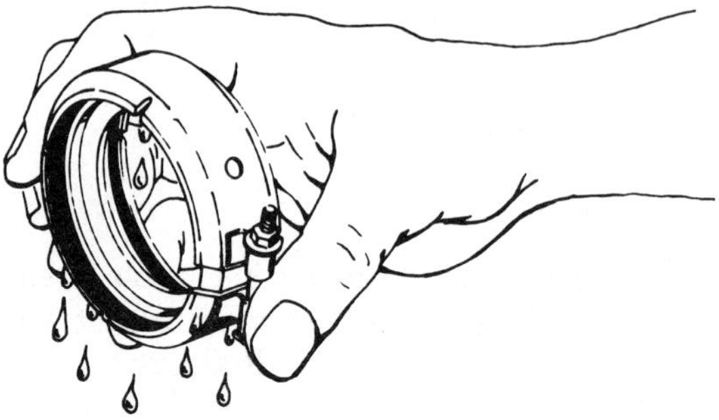

Fig. 4-21. A glass pipe coupler.

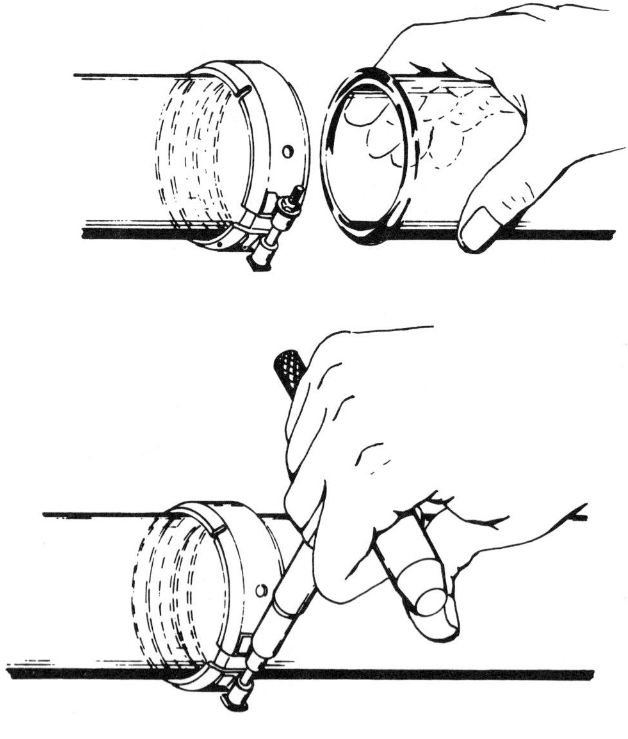

Fig. 4-22. How glass tubing is coupled.

Fewer joints are required, expansion joints are generally unnecessary, and fewer cleanouts and hangers are required. The pipe can be buried underground or under concrete floors. If the line should ever become plugged, one of three methods can be used to clear it:

1. Remove the plugged section of the line; the tubing is transparent, which makes the plugged section easy to detect.
2. Add chemicals to the system to dissolve the plugged area.
3. Use a special plastic-coated snake, which can be obtained from the pipe manufacturer.

Some glass chemical-waste line systems are ready packaged and can be purchased in kit form. The kit contains all the components necessary for any piping system in standard laboratory installations. Pre-engineered adjustable components eliminate the need to field fabricate short nipples. Fig. 4-20 illustrates a typical glass adjustable S-type drum trap.

The layout of glass piping is similar to that for other kinds of piping. For convenience, the dimensions of all pipe and fittings are measured to the center of each coupling. Special fittings can be made to order including odd lengths and elbows. Fig. 4-21 shows a pipe coupling dipped in water to make the connection easier. To close the coupling, use a standard ratchet wrench and socket, as shown in Fig. 4-22. Joints may be checked with a torque wrench to assure a positive leak-free compression joint.

CHAPTER 5

Blueprints

Building plans are printed in several different forms. The most common are blueprints, which are white lines on a blue background, and blue line drawings—blue lines on a white background. Architects and engineers have their individual style in preparing plans, but the basic principles are the same.

A building that starts out as an idea becomes a reality partly because the architect and the mechanical engineer know how to translate ideas into lines and symbols called *blueprints*. The architect and the engineer deserve only part of the credit for the finished project. It takes skilled mechanics, the building tradesmen, among whom are the plumber and the pipe fitter, to translate the language of blueprints into the finished project.

Two types of plans are used on most building projects: architectural plans and mechanical plans. Architectural plans show footing plans, floor plans, roof plans, details such as framing, and room finish schedules, which the plumber and pipe fitter should be able to read in order to fit the building piping into the available spaces.

Architectural plans are more rigid than mechanical plans; for instance, a footing must be placed exactly where the plans indicate, or the building may end up on someone else's property.

Mechanical plans may indicate that piping originates at a definite point, but unforeseen obstacles may cause it to deviate in some degree from the route shown on the plans. These obstacles are called "job conditions," and the plumber and the pipe fitter must be able to read and interpret the plans in order to overcome the job conditions and get the project finished in a workable manner.

Blueprints show room sizes, wall sizes, equipment layouts, pipe chases, room finish schedules, door sizes and so on. It is often necessary to jump back and forth between different sheets to find information because it is impossible to show all the details on one sheet. The following examples show how the plumber and the pipe fitter, using the information from different sections of the

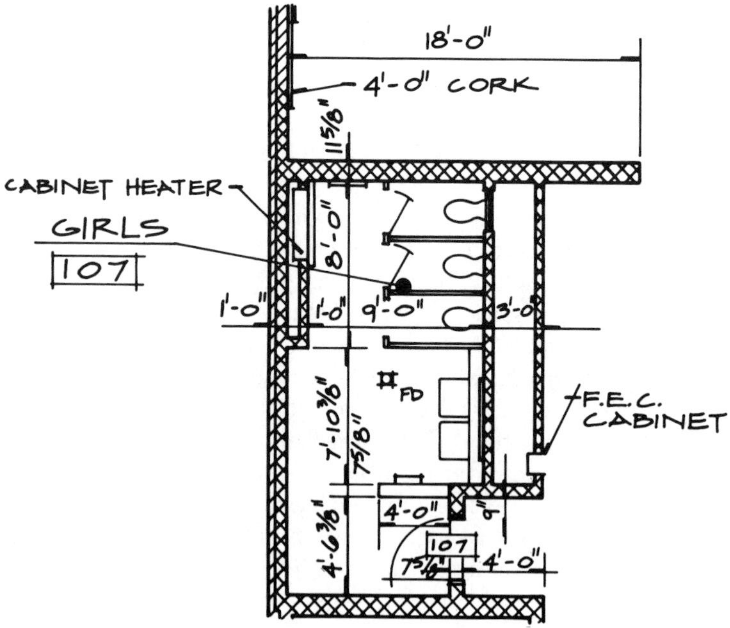

Fig. 5-1. Floor-plan scale: 1/8" = 1'-0".

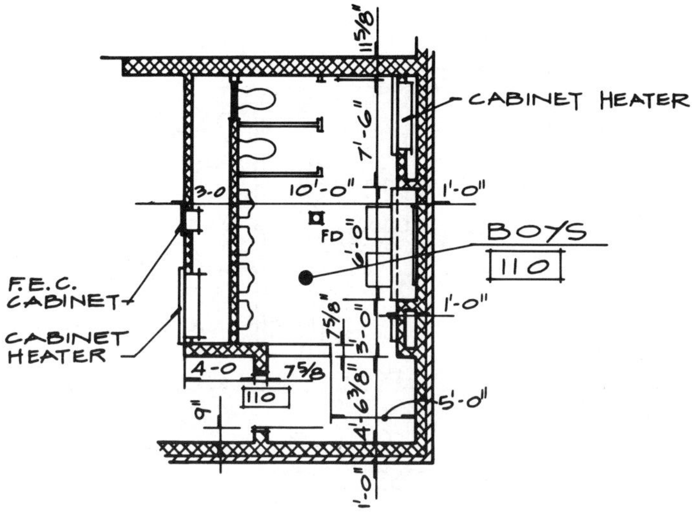

Fig. 5-2. Floor-plan scale: 1/8" = 1'-0".

plans, find the exact points at which the fixtures shown should be roughed.

Fig. 5-1 and 5-3 are plans of the same room, but the details shown in Fig. 5-2 clarify Fig. 5-1. Fig. 5-3 shows that the toilet partitions are 2'-8" cc (center to center) and the toilets will be centered in the partitions. The plumber knows that the toilets will be roughed in on 32" (2'-8") centers.

Fig. 5-1 shows the lavatories centered in a 7'-10 3/8" space, and Fig. 5-3 shows a 4'-0" mirror centered in this space. Using this information the plumber should rough in the lavatories within the 48" length of the mirror in order that the extreme edges of the lavatories do not project past the edges of the mirror when the fixtures are installed.

Fig. 5-1 indicates a cabinet heater is to be installed in the wall opposite the water closets. The detail drawing, Fig. 5-3, shows that the cabinet heater is to be centered on this wall. The location shown in Fig. 5-3 would take preference.

Both Fig. 5-2 and 5-4 show the four urinals centered in a space but the length of the wall is not given, it must be worked out. Starting from the inside of the 1'-0" wall, the room measures:

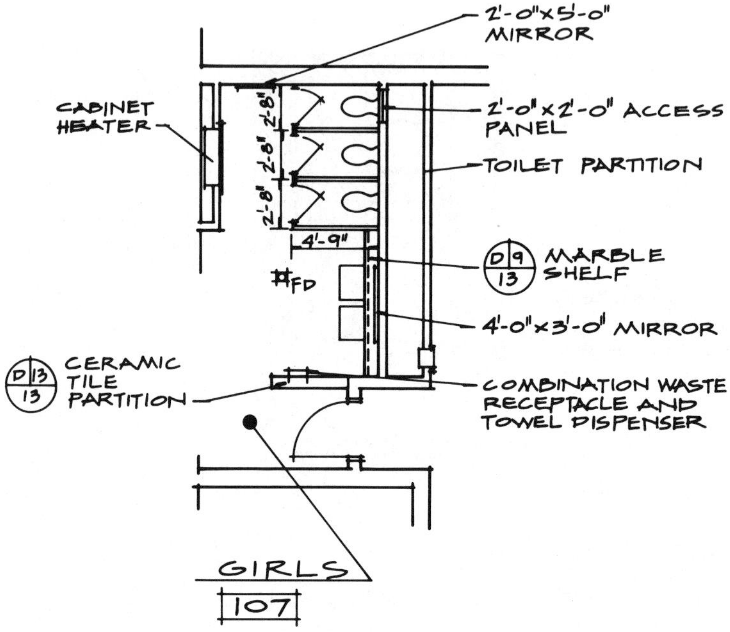

Fig. 5-3. Equipment-layout scale: $\frac{1}{8}''$ = 1′-0″.

$$\begin{array}{r} 4'\text{-}6\tfrac{3}{8}'' \\ +\ 3'\text{-}0\ '' \\ +\ 6'\text{-}0\ '' \\ +\ 7'\text{-}6\ '' \\ \hline 20'\text{-}12\tfrac{3}{8}'' \end{array} = 21'\text{-}0\tfrac{3}{8}''$$

The detail in Fig. 5-4 shows the toilet partitions are 2′-8″.

$$\begin{array}{r} 2'\text{-}8'' \\ +\ 2'\text{-}8'' \\ \hline 5'\text{-}4'' \end{array}$$

Fig. 5-2 shows the wall at the left end of the urinals to be 5′-2″ from the inside of the toilet room wall.

$$\begin{array}{r} 5'\text{-}2'' \\ +\ 5'\text{-}4'' \\ \hline 10'\text{-}6'' \end{array}$$

The room length is

$$\begin{array}{r} 21'\text{-}0\tfrac{3}{8}'' \\ -\ 10'\text{-}6'' \\ \hline 10'\text{-}6\tfrac{3}{8}'' \end{array}$$

The length of the wall on which the urinals are to be centered is 10′-6⅜″. The rough-in sheet for the urinals would show exactly how the fixtures should be centered in the space.

A symbol such as (D9/13) means that a detail, D9, will be found on sheet 13 of the plans. This type of symbol shows the detail number and the sheet on which it will be found.

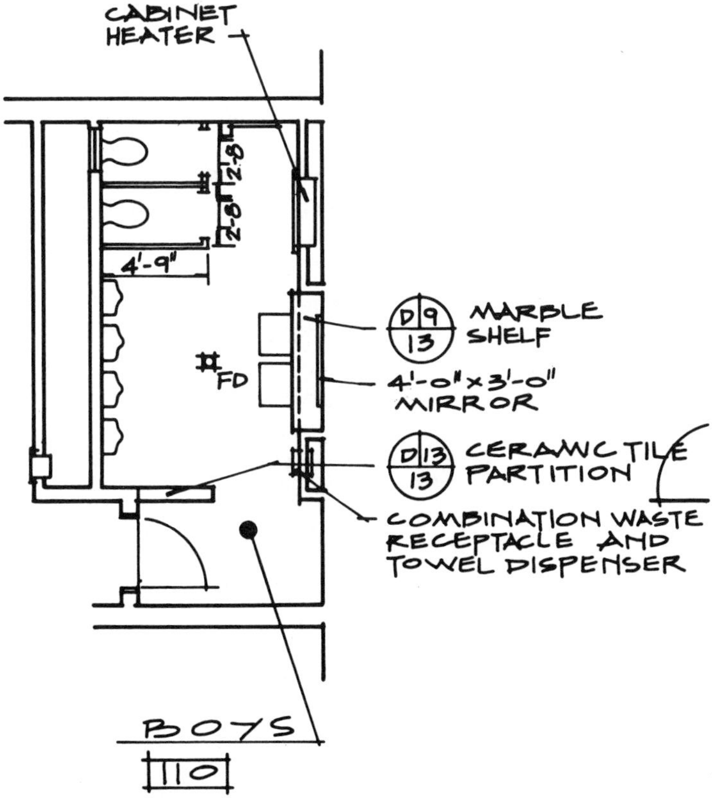

Fig. 5-4. Equipment-layout scale: ⅛″ = 1′-0″.

The following abbreviations are commonly used to indicate fire safety equipment:

F.E.C.—Fire Extinguisher Cabinet
F.H.C.—Fire Hose Cabinet

Common symbols used to identify pipe fittings and valves are shown in Fig. 5-5, 5-6, and 5-7.

MECHANICAL SYMBOLS

Fig. 5-8 is a section of a mechanical plan showing a male and female bathroom. The rooms are back to back, and the piping for these rooms is concealed in a pipe chase between the two rooms. The mechanical plans indicate that soil, wastes, vent, and hot- and cold-water piping are needed for these fixtures; but it is up to the plumber to figure how to get the necessary piping roughed in. Mechanical engineers usually make isometric drawings and include them in the plans. Fig. 5-9 is an isometric drawing of the piping shown is Fig. 5-8. A good isometric drawing will show almost every fitting needed to rough in the piping. Fig. 5-9A shows the soil, waste, and vent piping. Fig. 9B shows the cold-water piping. Fig. 5-9C shows the hot-water piping.

Building plans do not show the exact location of the fixtures or the soil, waste, vent, or water piping needed for the fixtures. This information is obtained from the roughing-in drawings.

ELEVATIONS

In order to read and understand blueprints it is necessary to have a working knowledge of elevations. The elevations of various parts of a building are shown on the plans. The *bench mark* is the starting point for working out the different elevations. A bench mark may be an iron rod driven into the ground in a protected location; it may be a nail driven into a tree; however the bench mark is established, it will be used as the reference point for elevations until some permanent point such as the top of a footing or a finished floor is available for a reference point. Any number can be used to indicate the different elevations of a building

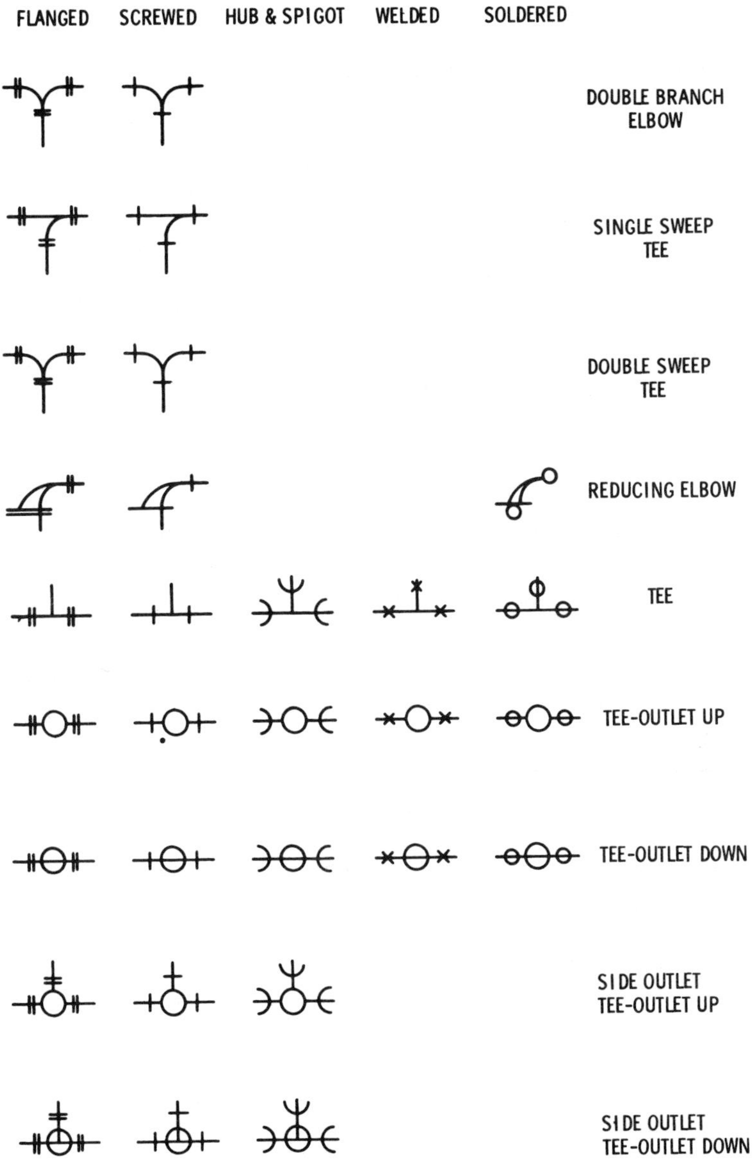

Fig. 5-5. Symbols for pipe fittings and valves.

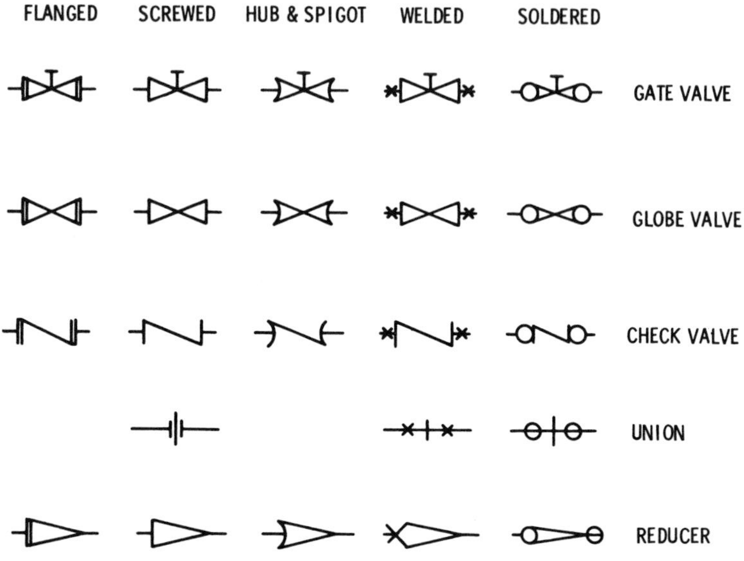

Fig. 5-6. Symbols for pipe fittings and valves.

project; quite often the number used is the approximate number of feet above sea level of the building site. A building site near the seacoast, at or near sea level, may use the number 100.00 as a reference point for elevations; while the number 750.00 might be used in land at that approximate height above sea level. In any case, the actual number used is not important; elevations higher than the bench mark are indicated by a higher number. If the finished first-floor elevation is shown as 750.00 and the second floor is shown as 760.00, it shows that the second floor is 10 ft. above the first floor. If the basement elevation is shown as 738.00, the basement will be 12 ft. lower than the finished first floor. The important thing to remember when working with elevations is that a number larger than the bench mark is *above* the bench mark; a smaller or lower number indicates a point *below* the bench mark, as shown in Fig. 5-10.

As shown in Fig. 5-10, the bench mark is 747.50. The finished first-floor elevation is 750.00. The first floor is 2.50 (2½) ft. *above* the bench mark. The basement floor at an elevation of 738.00 is

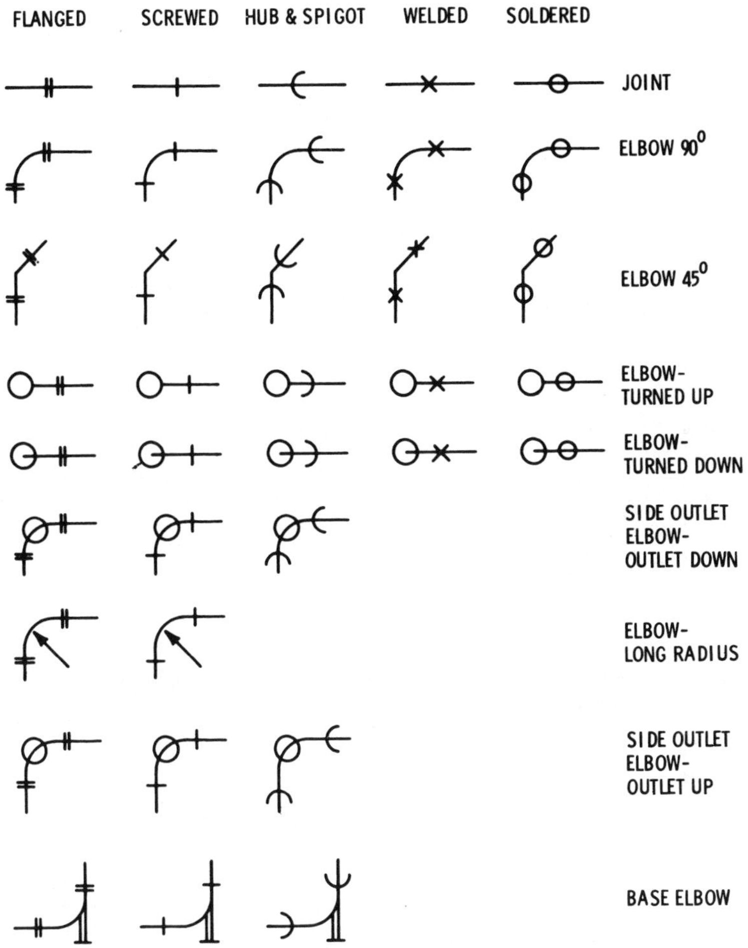

Fig. 5.7. Symbols for pipe fittings and valves.

9½ ft. (9.50) *below* the bench mark and 12 ft. below the finished first floor.

The plumber and pipe fitter should be familiar with the use of the instrument level (Fig. 5-11). An instrument level is a telescope, mounted on a base equipped with leveling screws, the base is screwed onto a tripod (Fig. 5-12) when the instrument is in use. The instrument level can be rotated a full 360° horizontally and

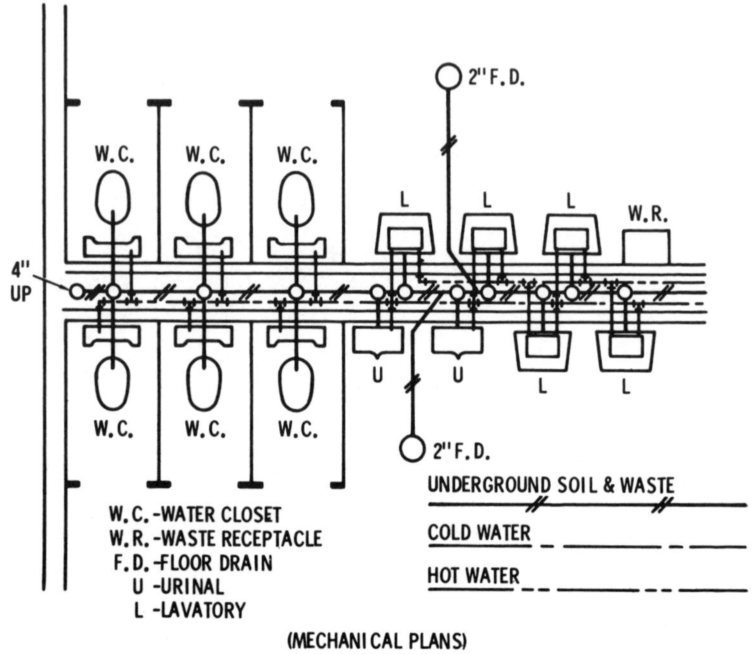

Fig. 5-8. Section of a mechanical plan showing male and female bathrooms.

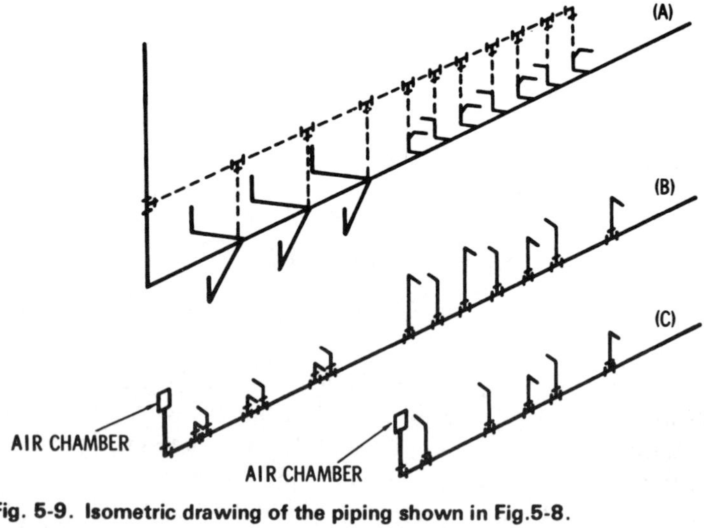

Fig. 5-9. Isometric drawing of the piping shown in Fig.5-8.

ROOF ELEVATION	780.00
3RD. FLOOR ELEV.	770.00
2ND. FLOOR ELEV.	760.00
1ST. FLOOR ELEV.	750.00
BENCH MARK 747.50	
BASEMENT ELEV.	738.00

Fig. 5-10. A typical example of elevations.

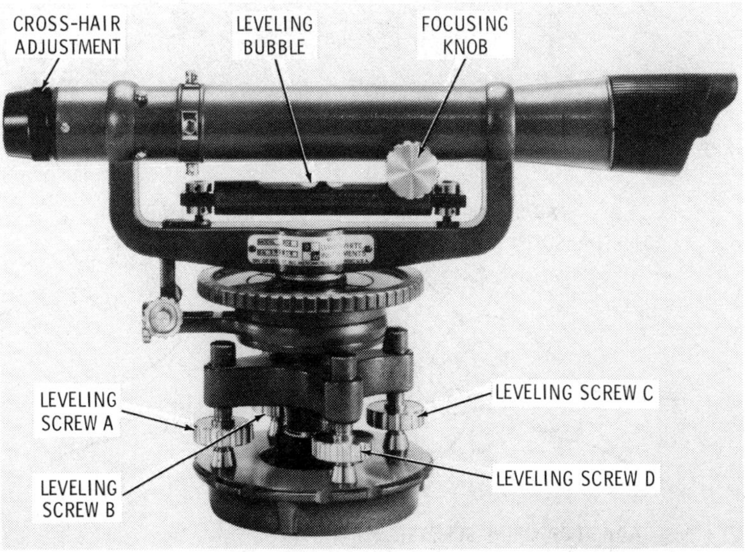

Courtesy David White Instruments

Fig. 5-11. An instrument level can be used to establish elevations.

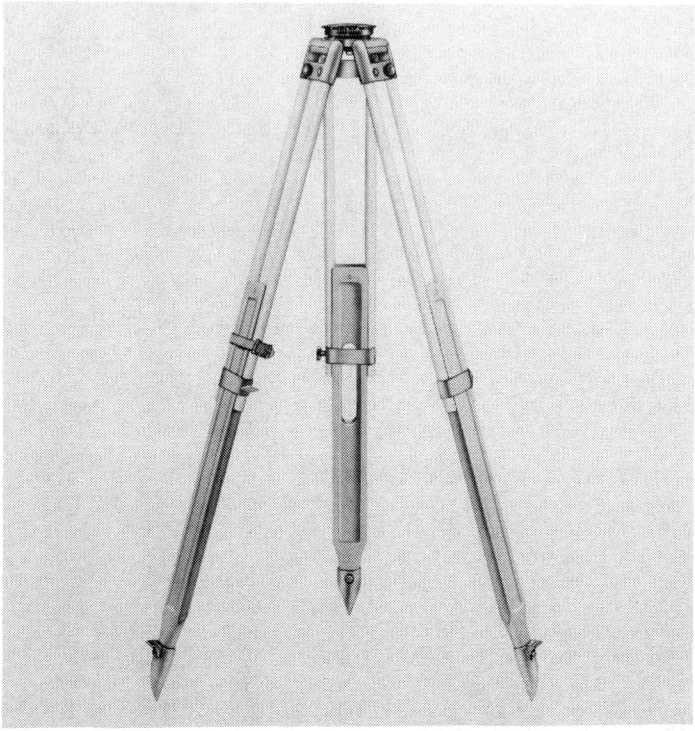

Courtesy David White Instruments

Fig. 5-12. Tripod on which instrument level can be mounted.

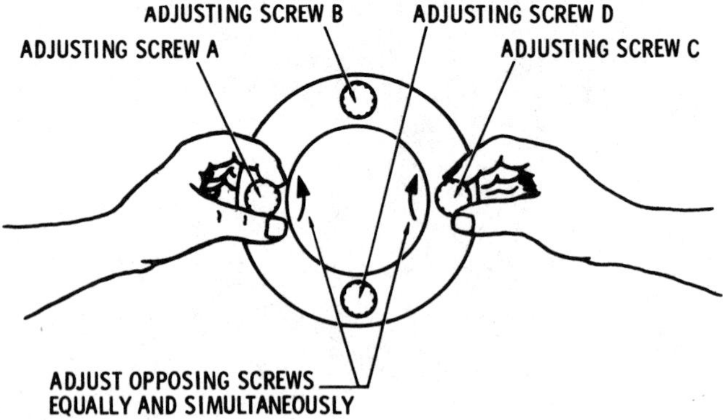

Fig. 5-13. Leveling adjustments.

when properly adjusted the level-indicating bubble will show that the instrument is level at any point of the compass. To set up the instrument the tripod should be opened and the legs spread apart so that the top of the tripod is approximately level with the legs of the tripod pressed firmly into the ground. The instrument level is then screwed onto the tripod and swung to a position lined up directly over adjusting screws A and C. Both screws should be turned equally and simultaneously, as shown in Fig. 5-13, to center the leveling bubble. Turning both screws "in" moves the bubble to the right. Turning both screws "out" moves the bubble to the left. When the bubble is centered, swing the instrument to line up directly over screws B and D and repeat the process. When the bubble is centered, return the instrument to line up over A and C again and re-adjust the screws to again level the instrument. The leveling process may have to be repeated several times until the instrument is in level position when pointed in any direction. A word of caution: Do not *overtighten* the adjusting screws. Overtightening can damage the instrument.

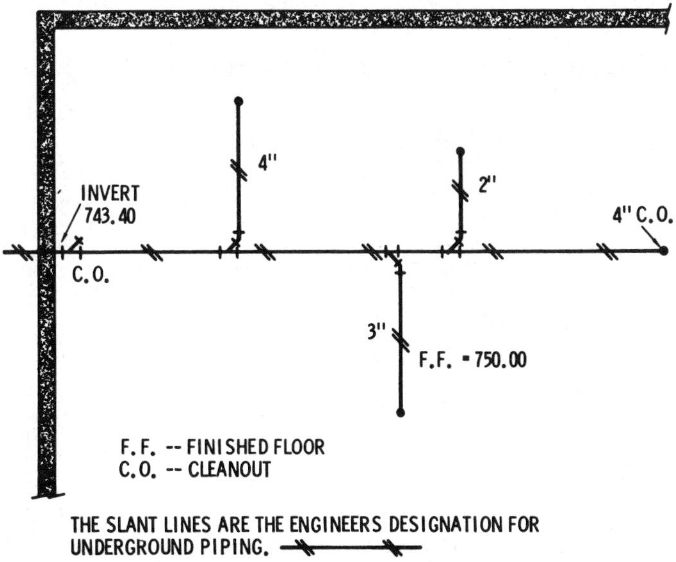

Fig. 5-14. A partial mechanical plan of a building.

Using an Instrument Level

Fig. 5-14 shows a partial mechanical plan of a building. The invert (or bottom inside) of the building drain is shown as 743.40. Although the finished floor elevation is shown as 750.00, this figure cannot be used to find the invert elevation because the construction is just starting. The plumber must set up an instrument level and set the building drain to the invert shown.

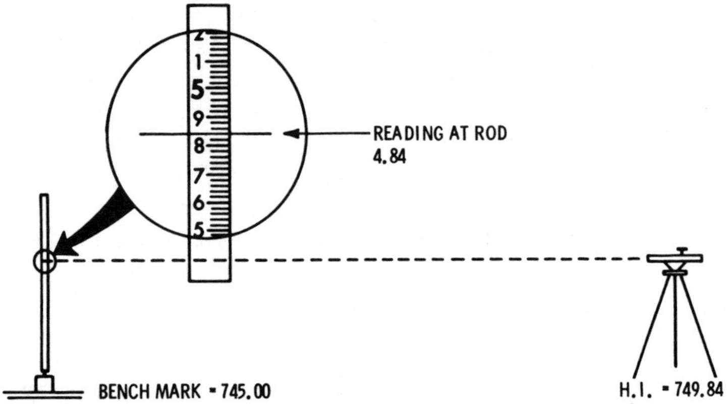

Fig. 5-15. Determining the height of an instrument.

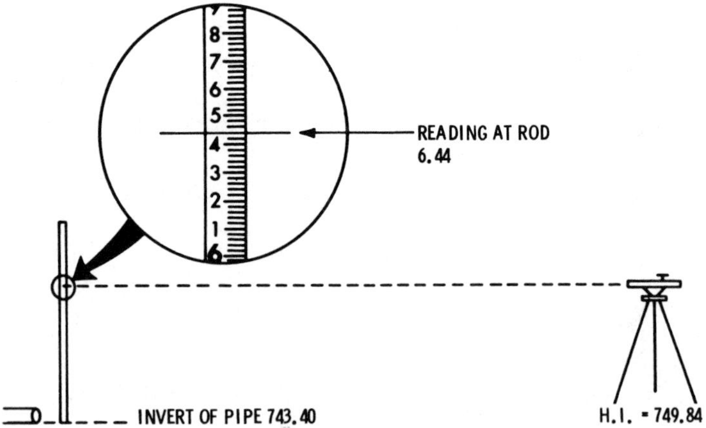

Fig. 5-16. Finding the building drain elevation.

As shown in Fig. 5-15, with the rod held on the bench mark a reading of 4.84 will be made. The bench mark is at 745.00; this figure is added to the 4.84 reading, to establish the H.I. (height of instrument). This figure, 749.84, is the actual elevation of the instrument. If the invert elevation is subtracted from the H.I., the result, 6.44, is the number the plumber should read when the building drain is at the correct elevation, as shown in Fig. 5-16:

$$\begin{array}{r} 749.84 \\ -\ 743.40 \\ \hline 6.44 \end{array}$$

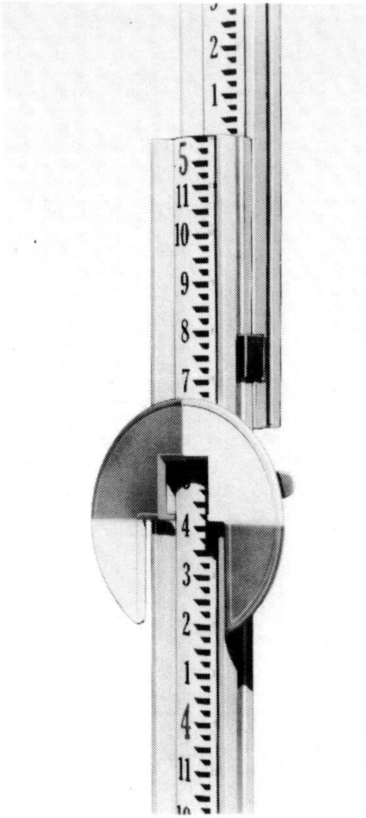

Fig. 5-17. A close-up view of a typical engineer's rod.

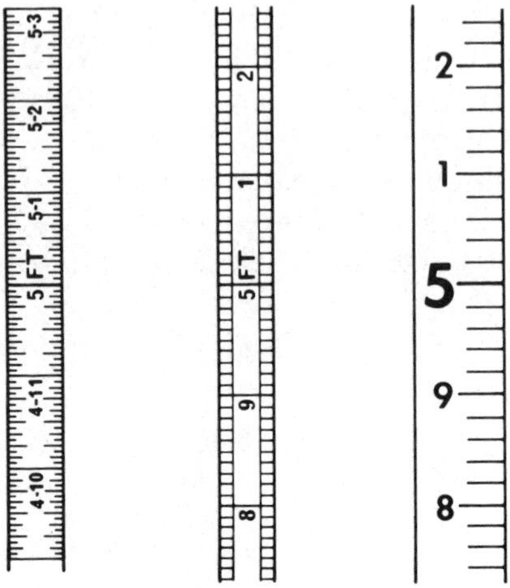

Fig. 5-18. A standard rule, an engineer's rule, and an engineer's rod.

HELPFUL CONVERSIONS AND EQUIVALENTS

Approximate Conversions from Metric Measures

Symbol	When You Know	Multiply by	To find	Symbol
		LENGTH		
mm	millimeters	0.04	inches	in
cm	centimeters	0.4	inches	in
m	meters	3.3	feet	ft
m	meters	1.1	yards	yd
km	kilometers	0.6	miles	mi

Approximate Conversions to Metric Measures

Symbol	When You Know	Multiply by	To Find	Symbol
		LENGTH		
in	inches	*2.5	centimeters	cm
ft	feet	30	centimeters	cm
yd	yards	0.9	meters	m
mi	miles	1.6	kilometers	km

Fractional Inches	1/64	1/32	1/16	1/8	1/4	1/2	3/4
Decimal Inches	.016	.031	.063	.125	.25	.50	.75

Courtesy David White Instruments.

Fig. 5-19. A metric standard conversion guide.

It is very important when holding a rod to obtain a reading that the rod be held straight. If the rod is leaning toward the instrument, the reading will be high; if leaning away, the reading will be low. Fig. 5-17 shows a close-up view of the rod.

An engineer's 6 ft. rule shows feet and inches on one side and measurements in 10ths and 100ths on the other side. As shown in Fig. 5-18A, using an engineers rule a measurement of 4′-10⅞″ can be converted to 4.91 by reading the 4′-10⅞″ on one side and turning the rule over and reading the 4.91 on the other side, Fig. 5-18B. This type of rule is very useful when working with prints which give some elevations or measurements in ft. and in. and other measurements in ft., 10ths and 100ths. Conversion guides such as Fig. 5-19 are helpful in converting measurements to or from metric measures.

CHAPTER 6

Pipe Fitting

The term *pipe fitting* describes the operations that must be performed in installing a pipe system made up of pipe fittings. These operations consist of:

1. Pipe cutting.
2. Pipe threading.
3. Pipe tapping.
4. Pipe bending.
5. Assembling.

The mechanic who performs the work of pipe fitting is called *pipe fitter* (sometimes a *steam fitter* because the work is often connected with steam installations). Considerable experience is necessary to become a good pipe fitter.

PIPE CUTTING

Steel received from the manufacturer comes in lengths varying from 20 to 22 ft. (usually 21 ft.), which makes it necessary to cut it to the proper length. This may be done with a hacksaw or a pipe cutter, the pipe in either case being held in a pipe vise such as the one shown in Fig. 6-1. Care should be taken in securing the pipe in the vise (especially when threading) that the jaws hold the pipe sufficiently firm to prevent slipping, but the clamp screw should not be tightened so tightly that the jaw teeth will dig excessively into the pipe.

Most pipe cutters consist of a hook-shaped frame on whose stem a slide is moved by a screw. One or more cutting discs or wheels are mounted on the slide and/or frame, and forced into the metal as the tool is rotated around the pipe. Pipe cutters may be classified as:

1. Wheel.
2. Combined wheel and roller.

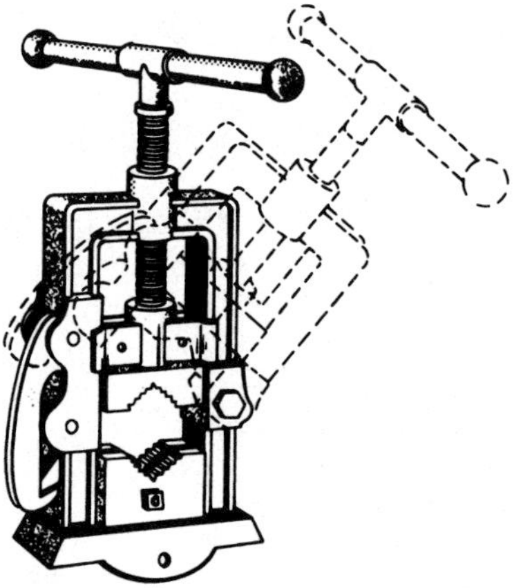

Fig. 6-1. A pipe vise.

A combined wheel-and-roller is shown in Fig. 6-2. This particular model can be converted to a three-wheel cutter by replacing the two rollers with cutter wheels. Another type of pipe cutter is shown in Fig. 6-3. This cutter is used to cut 2 in. to 6 in. clay, cast-iron, asbestos-cement, and 4 in. water main pipe.

Pipe can be cut more quickly and accurately with a pipe cutter than with a hacksaw. The hacksaw does not leave as much of a burr on the pipe as a cutter, but the greater speed and accuracy of the pipe cutter dictates its use in nearly all pipe-fitting work.

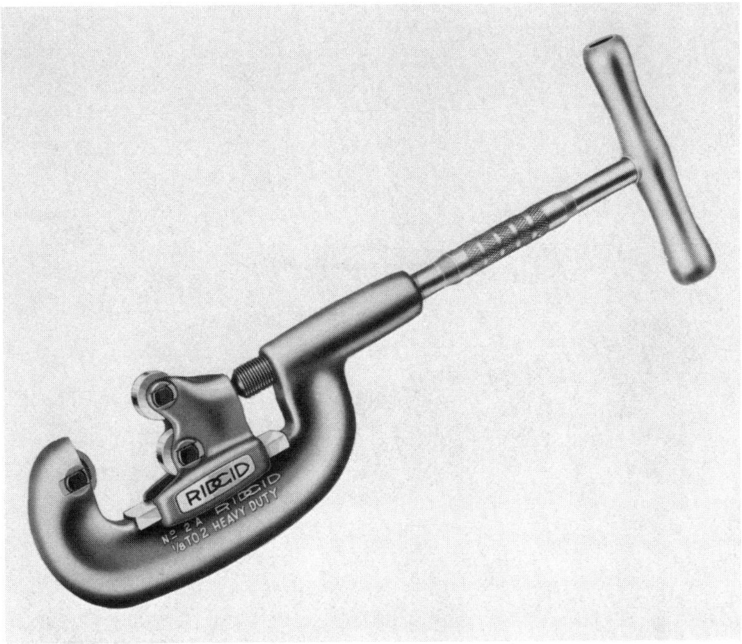

Courtesy Ridge Tool Co.

Fig. 6-2. A single-wheel pipe cutter.

The single-wheel cutter is best for all-around work, but there are times when a three-wheel cutter is necessary, as shown in Fig. 6-4. Here, it shows the impossibility of cutting a pipe in a close-fitting space with a wheel-and-roller cutter. A 360° rotation of the entire cutter is necessary to cut the pipe. However, with a three-wheel cutter, a rotation of slightly more than 120° is all that is necessary to cut the pipe. The cutter wheels on either type are easily removed and renewed when they become dull or nicked.

A little more care must be taken when starting a cut with a three-wheel cutter than with the roller type to make certain the cut is straight. In addition, the three-wheel type leaves more of an outside burr. This burr, as well as the burr on the inside of the pipe, must be removed on every cut to avoid future trouble with clogged pipes. The outside burr will also interfere with starting the pipe threader, so it should be removed with a file or rasp before attempting the threading operation. A convenient way to remove

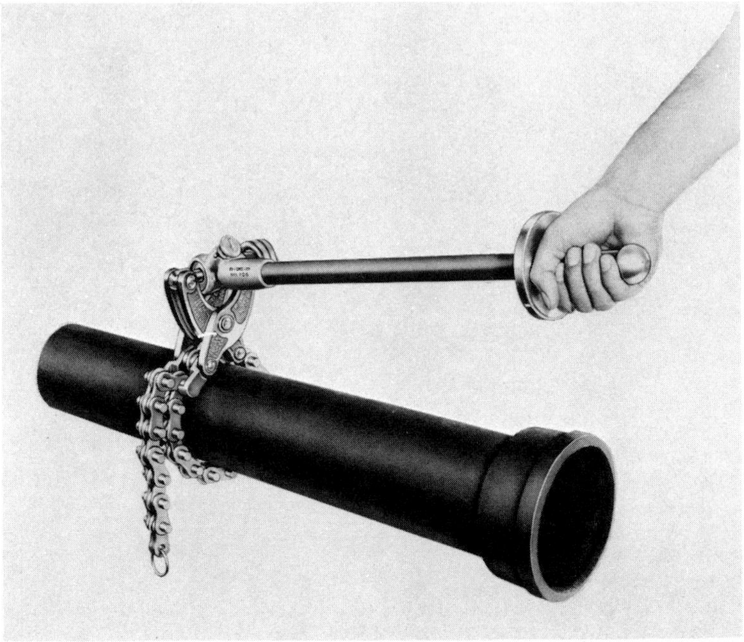

Courtesy Ridge Tool Co.

Fig. 6-3. A soil-pipe cutter. This cutter weighs 18 lbs. and has the power to cut 2- to 6-in. clay, cast-iron, and asbestos soil pipe, and 4-in. water main.

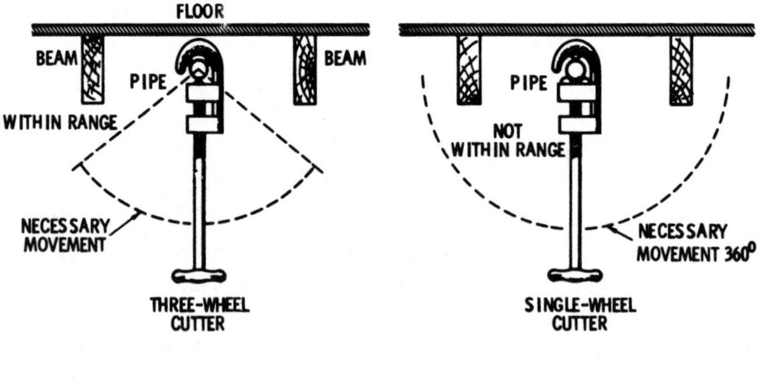

Fig. 6-4. The difference between a three-wheel and a single-wheel pipe cutter when used in a confined space.

the internal burr is to use a reamer designed for use in a hand brace, or to use a reamer similar to the types shown in Fig. 6-5.

PIPE THREADING

Having cut the pipe to proper length, filed off the outer shoulder, and reamed out the burr, it is now ready for the thread-

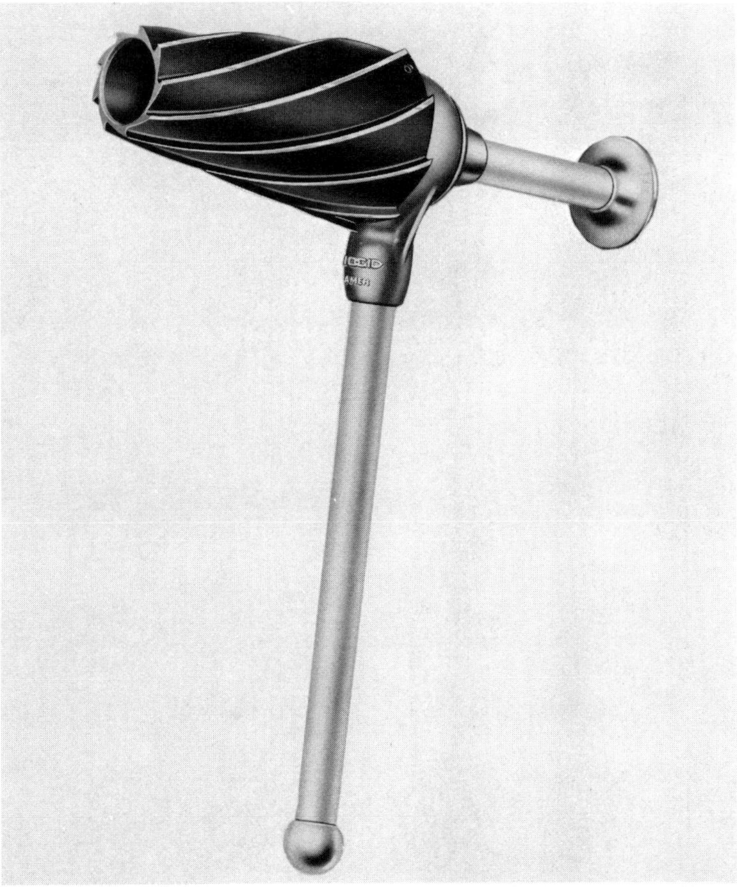

Courtesy Ridge Tool Co.

Fig. 6-5. Reamers used to remove the burr from the inside diameter of a pipe.

ing operation. The threads may be cut on the pipe ends by means of:

1. Hand stock and dies.
2. Pipe-threading machines.

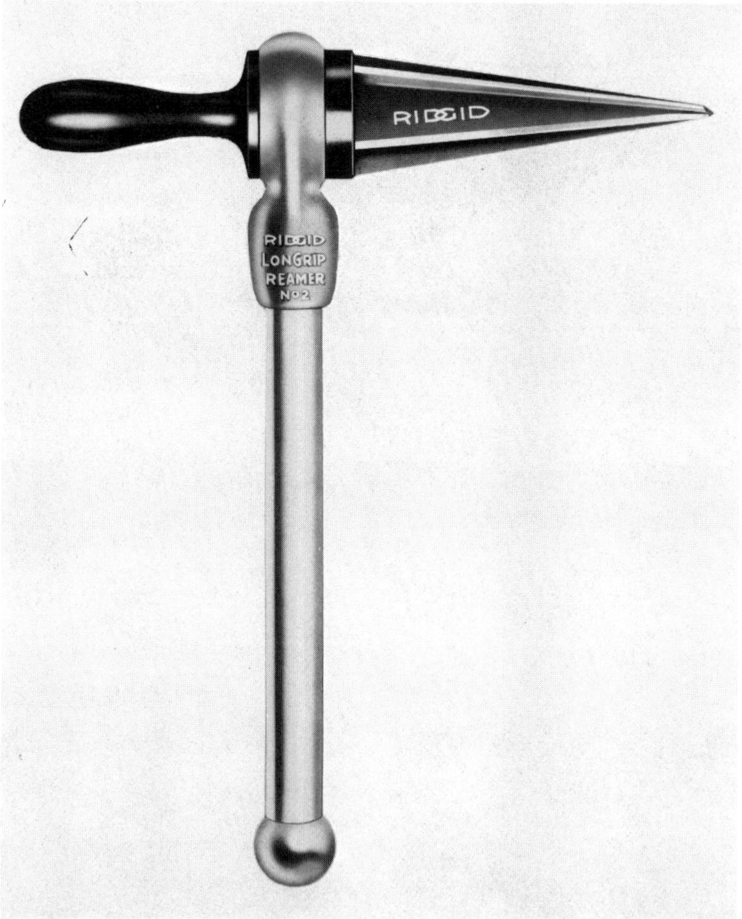

Courtesy Ridge Tool Co.

Fig. 6-5. Reamers used to remove the burr from the inside diameter of a pipe (cont'd).

Hand-Operated Threaders

The hand stock and dies are portable, and are generally used for small jobs, especially for threading pipe of the smaller sizes. Threading machines are for use where a large amount of threading is done.

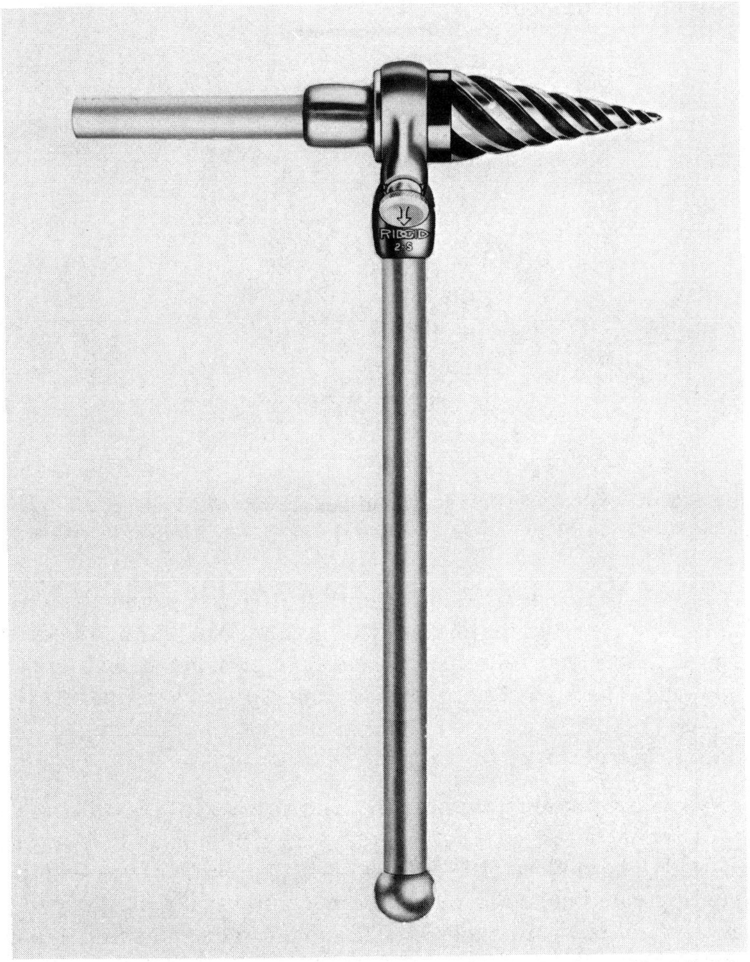

Courtesy Ridge Tool Co.

Fig. 6-5. Reamers used to remove the burr from the inside diameter of a pipe (cont'd).

Fig. 6-6 shows a popular hand-operated stock and dies combining threading facilities for three different pipe sizes in one unit. This threader is for pipe size ⅜ in. through ¾ in. and is instantly ready to cut any one of three (⅜″ - ½″ - ¾″) sizes of threads. Each die set is securely locked in place and includes a guide to ensure straight and true threads. The dies can be reversed in the holder for close-to-wall threading.

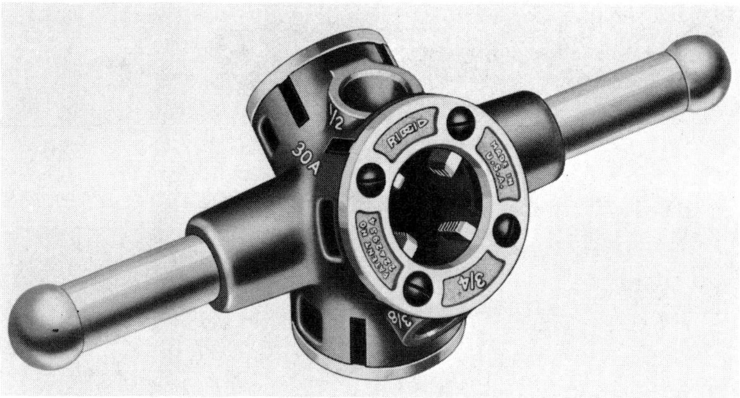

Courtesy Ridge Tool Co.

Fig. 6-6. A combination pipe stock and dies capable of threading ⅜-, ½-, and ¾-in. pipe. Combinations for other pipe sizes are available.

Another hand-operated threader (Fig. 6-7) has a ratchet handle and will accept dies for threading pipe from ¼ in. through 1 ¼ in. The die heads snap in from either side and push out easily for fast changing. The ratchet permits threading pipe in close quarters.

Power Threaders

Where large amounts of pipe are to be threaded a power threading machine (Fig. 6-8) offers a great saving in time and labor. In addition, the quality of the thread is usually better than with a hand-operated unit. The most versatile, and usually most expensive, type of power threader is a self-contained unit powered by an electric motor. These threaders have a built-in pipe cutter, a reamer, and a pump to circulate threading oil and direct it to the proper place where it floods the threading area. The dies for

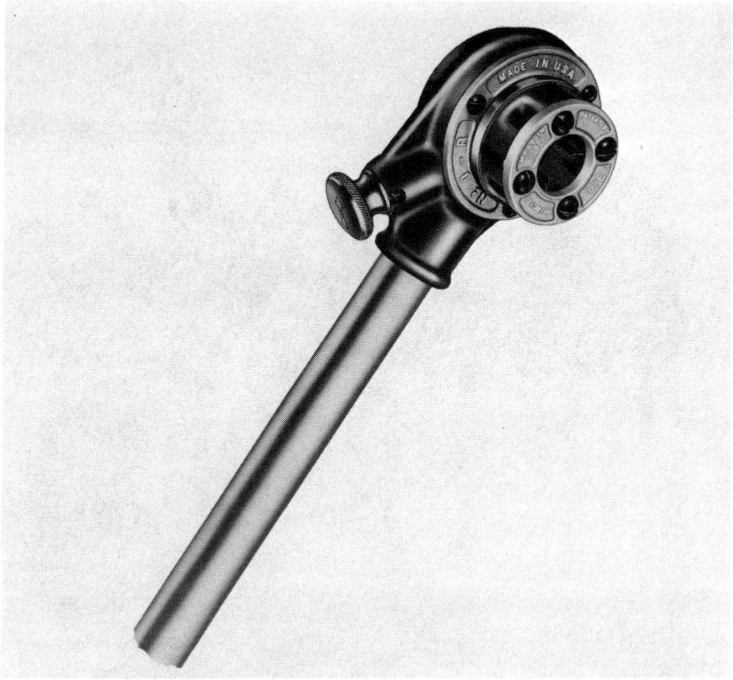

Courtesy Ridge Tool Co.

Fig. 6-7. Ratchet-type pipe dies feature interchangeable dies plus a ratchet action for ease of operation.

different sizes of pipes are instantly interchangeable, making the machine ideal for all types of plumbing and pipe fitting. A typical pipe die for use with a power threading machine is shown in Fig. 6-9.

Another type of threading die worthy of note is the geared pipe threader shown in Fig. 6-10. This threader can be used for either hand or power threading. The die-size selector plate is easily and quickly set, and locks at the desired size. This device also features an adjustment for tapered, straight, oversize, or undersize threads.

A good grade of thread-cutting oil should always be used when threading pipe. Thread-cutting oil makes the threading operation much easier, protects and maintains the dies in good condition, and helps produce clean-cut and accurate pipe threads.

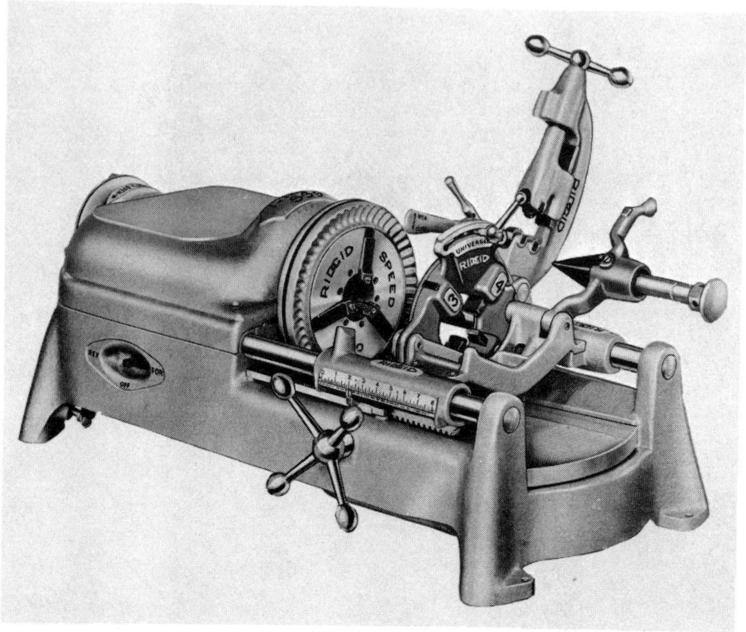

Courtesy Ridge Tool Co.

Fig. 6-8. A power threading machine. This unit has dies to fit all popular pipe sizes and permits cutting to length and reaming.

Threading Operation

When using hand threaders, press the dies firmly against the pipe until they take hold. Apply plenty of thread-cutting oil. After a few turns, blow out the chips and apply more oil. This procedure should be repeated two or three times before completing the cut. When completed, blow out as many chips as possible and back off the die. Avoid frequent reversals during the threading operation as this results in damaged threads and possible damage to the dies.

Flat Threads

Pipe is often discarded because of the presence of flat or improper threads. Most of this pipe can be used without sacrificing the quality of the installation. It has been found that the entire thread must be flat in order to cause a leak. Actually, only a very

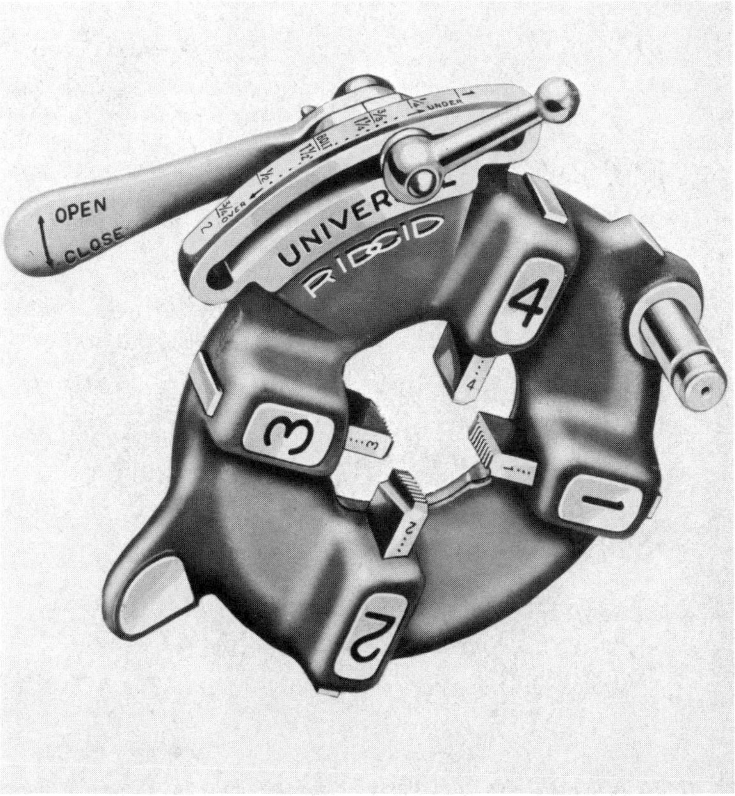

Courtesy Ridge Tool Co.

Fig. 6-9. Quick-opening pipe dies for use with a die-hard adapter or a power threading machine.

few perfect threads will produce a tight and leakproof joint. Pipe-joint compound, thread seal, or Teflon should always be used because it will make the installation easier and ensures trouble-free joints.

PIPE TAPPING

It is often necessary to cut internal threads in pipes and fittings such as in pipe headers, lubricator connections, etc. This process is called *tapping* and involves one or more of the following steps:

Courtesy Ridge Tool Co.

Fig. 6-10. A geared pipe threader that can be used for power or hand threading.

1. Drilling holes to the correct diameter.
2. Reaming.
3. Cutting the internal threads with a tap.

It is necessary to know the correct size of hole required for each size of pipe tap. Table 6-1 lists the correct drill size to use.

In drilling a hole to be tapped, care should be taken to guide the drill correctly so that the hole will be aligned properly. In tapping, the tap must not be turned with too much force or it will break. This is especially true of small taps. If the tap does not turn reasonably easy, work it back and forth and back it off occasionally to remove the chips. Always use a generous amount of thread-cutting oil. A typical pipe tap is shown in Fig. 6-11.

Table 6-1. Drill Sizes for Standard Pipe Taps*

Pipe Size	Threads per Inch	Drill Size
1/8	27	11/32
1/4	18	7/16
3/8	18	19/32
1/2	14	23/32
3/4	14	15/16
1	11 1/2	15/32
1 1/4	11 1/2	1 1/2
1 1/2	11 1/2	1 23/32
2	11 1/2	2 3/16
2 1/2	8	2 5/8
3	8	3 1/4
3 1/2	8	3 3/4
4	8	4 1/4
4 1/2	8	4 3/4
5	8	5 5/16
6	8	6 3/8

*To secure the best results, the hole should be reamed before tapping with a reamer having a taper of 3/4" per foot.

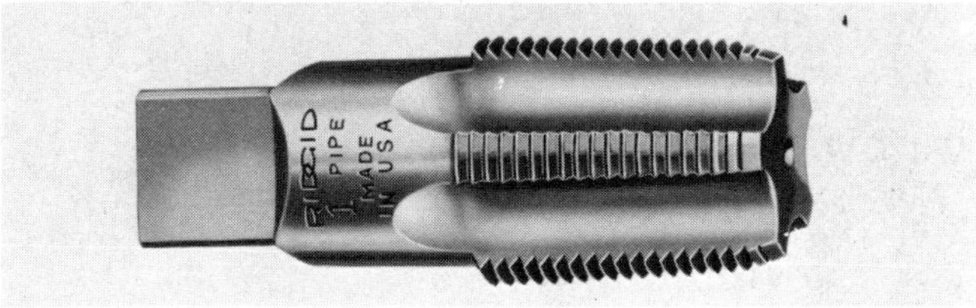

Courtesy Ridge Tool Co.

Fig. 6-11. A pipe die for cutting internal threads.

CALCULATING OFFSETS

In pipe fitting, the term *offset* may be defined as *a change of direction (other than 90°) in a pipe that brings one part out of, but parallel with, the line of another pipe*. An example of this is illustrated in Fig. 6-12, where it is necessary to change the position of pipe line, *L*, to a parallel position, *F*, in order to avoid some obstruction, such as wall *E*. When the two lines, *L* and *F*, are to be fitted with elbows having an angle of other than 90°, the pipe fitter

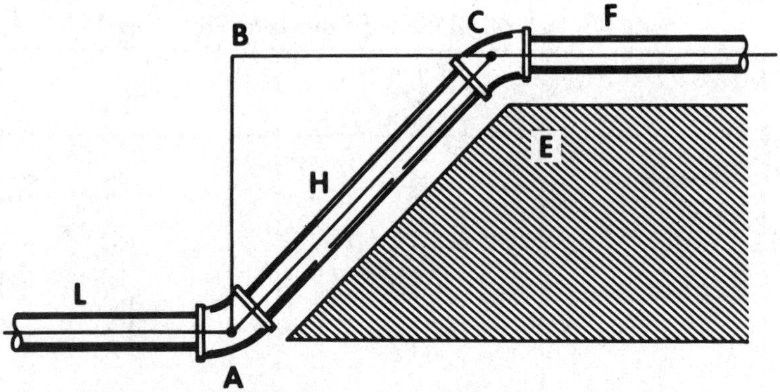

Fig. 6-12. A pipe line connected with 45° elbows showing offset and the method of finding the length of the connecting pipe.

must find the length of the pipe *H* connecting the two elbows, *A* and *C*. The distance *BC* must also be determined in order to fix the point *A*, so that elbows *A* and *C* will be in alignment. There are several methods of solving this problem; four of these methods are given here.

Method 1

In the triangle *ABC*

$$(AC)^2 = (AB)^2 + (BC)^2$$

from which

$$AC = \sqrt{(AB)^2 + (BC)^2}$$

Example—If the distance between pipe lines *L* and *F* in Fig. 6-14 is 20 inches (offset *AB*), what length of pipe *H* is required to connect with the 45° elbows *A* and *C*? When 45° elbows are used, both offsets are equal. Thus, substituting in the equation,

$$AC = \sqrt{20^2 + 20^2} = \sqrt{800} = 28.28 \text{ inches}$$

The length of the pipe just calculated does not allow for the projections of the elbows. This must be taken into account, as shown in Fig. 6-13.

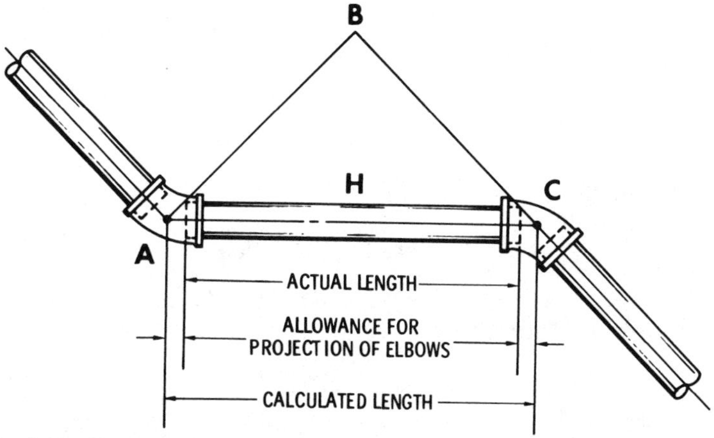

Fig. 6-13. The calculated and actual length of connecting pipe with elbows other than 90°.

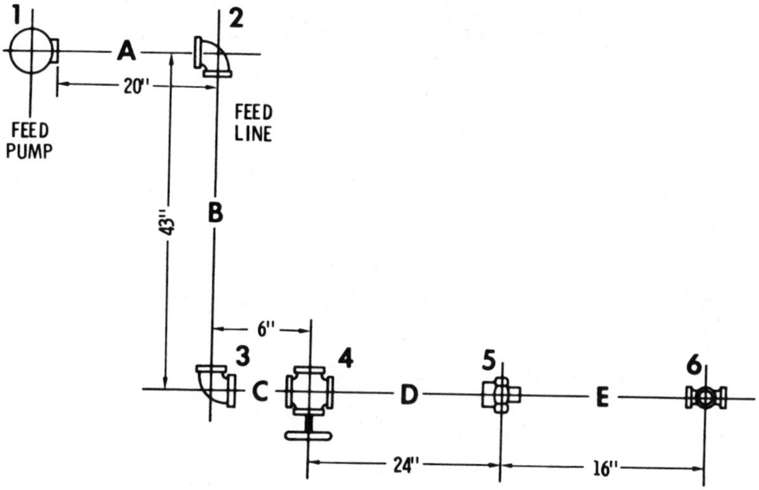

Fig. 6-14. A centerline sketch with dimensions for pipe fitting entirely by measurement.

Method 2

The following rule will be found convenient in determining the length of the pipe between 45° elbows.

Rule—*For each inch of offset, add* 53/128 *of an inch, and the result will be the lengths between centers of the elbows.*

Example—Calculate the length *AC* (Fig. 6-14) by preceding rule.

$$20 \times \frac{53}{128} = \frac{1060}{128} = 8\,{}^{9}\!/_{32}$$

Adding this to the offset,

$$2 + 8\,{}^{9}\!/_{32} = 28\,{}^{9}\!/_{32} \text{ inches}$$

This is the calculated length; deduct the allowance for the projection of the elbows, as in Fig. 6-15.

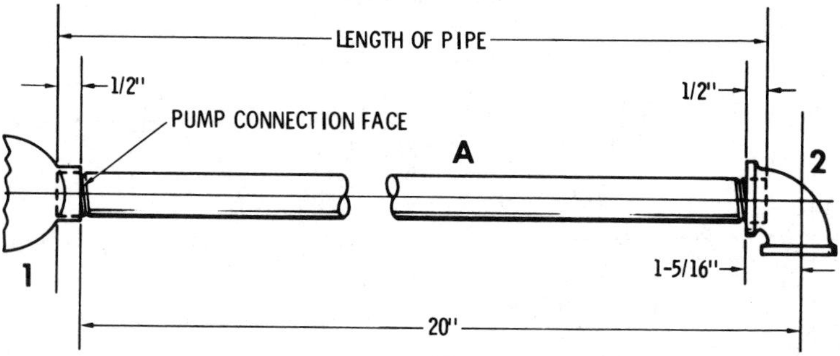

Fig. 6.15. Detail of a portion of the feedline shown in Fig. 6-14 illustrating the method of determining the exact length of pipe.

Method 3

Elbows are available with angles other than 45° and 90°. For instance, angles of 60°, 30°, 22½°, 11¼°, and 5⅝° are manufactured and sometimes encountered by the pipe fitter. When such elbows are used, the distance between their centers can be easily found by using the constants listed in Table 6-2.

Rule—*To find the length between centers of the elbows, multiply the offset by the constant for the elbows used.*

Referring to the figure in Table 6-2,

AC = offset AB × constant for AC

BC = offset AB × constant for AB

Table 6-2. Elbow Constants

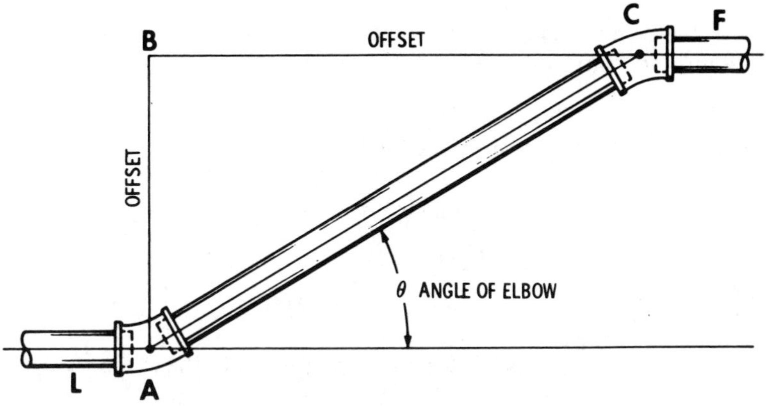

Elbow Angle	Elbow Centers (AC)	Offset (AB)
60°	1.15	0.58
45°	1.41	1.00
30°	2.00	1.73
$22\frac{1}{2}$°	2.61	2.41
$11\frac{1}{4}$°	5.12	5.02
$5\frac{5}{8}$°	10.20	10.15

Example—If the distance between pipe lines *L* and *F* (offset *AB*) is 20 inches, what is the length of offset *BC*, and the distance *AC* between the center of the elbows, if $22\frac{1}{2}$° elbows are used?

From the table, the constant for *AB*, for a $22\frac{1}{2}$° elbow, is 2.41. Substituting the values in the proper equation,

$$BC = 20 \times 2.61 = 52.2 \text{ inches}$$

For the distance *AC* between the elbows, the constant in the table is 2.61. Substituting the values in the proper equation,

$$AC = 20 \times 2.61 = 52.2 \text{ inches}$$

Method 4

Offsets may also be calculated by using trigonometry. The following examples illustrate the use of the tangent function.

Example—In Fig. 6-8 from Chapter 6 in Vol. 3, what is the length of offset *OB*?

$$\tan 60° = 1.7321$$

Length of offset *OB:*

$$1.7321 \times 8 = 13.86$$

Example—What is the length of offset *AB* in the same figure? From the table,

$$\tan 30° = 0.5774$$

Length of offset *AB:*

$$\begin{aligned} AB &= \tan 30° \times OB \\ &= 0.5774 \times 13.86 \\ &= 8 \text{ inches} \end{aligned}$$

ASSEMBLY

The pipe on large jobs is usually cut according to a sketch or working drawing and partly assembled at the shop. If no mistakes have been made in following the dimensions on the drawing, and if the drawings are correct, the pipe and fittings may be installed without difficulty—that is, the last joint will come together or *make up*. This last joint is either a union, a right-and-left, or a long screw joint, and if errors have been made in cutting the pipe, it will be difficult or impossible to make up this closing joint.

On small jobs, no sketch is necessary. The plumber proportions the pipe lengths mostly "by eye," taking occasional measurements where necessary during the progress of the work. It should be noted, that, with the great variety of fittings available, any pipe system may be arranged in a number of different ways. The proper selection of these fittings, and the general arrangement of the system so that it will be direct, simple, accessible for repairs, etc., is an index of the pipe fitter's ability.

Joint Compound

In making up screwed joints, a good joint compound is mandatory. This compound serves two useful purposes—it lubricates the joint, making the tightening process much easier, and also forms a seal to ensure a tight joint. Many different materials can be used as a joint compound, but the best is that manufactured for this purpose. Teflon tape, which is also available, is easy to use and works well and efficiently. If neither of these types of joint materials is available, white or red lead, or graphite, can be used with some success. Red lead will provide a tight joint, but has the disadvantage of making it difficult to unscrew the joint in case of future repairs.

In applying the joint compound, *it should be put on the male thread only.* If put on the female thread, some of the compound will lodge inside the pipe, forming an obstruction or contaminating the liquid, which will flow through the pipe. An old toothbrush or similar object provides a handy means for applying the compound.

Joint Makeup

When making up a joint, it should not be tightened too rapidly. The tightening process process produces heat (because of friction), which may expand the pipe enough that the proper number of turns cannot be made. After the joint has cooled, it may be loose enough to allow a leak. The approximate distance that different sizes of pipe should extend into fittings for a tight joint is listed on Table 6-3. This length of thread must be taken into account when cutting the pipe to fit in a given space.

Some pipe fitters, when called upon to install piping that will be subjected to pressures of 200 to 300 psi (such as some steam lines), feel that they are assuming a heavy responsiblity. If asked to make an installation that will carry a pressure of 1000 psi, they might feel the responsibility is too great. Actually, joints tight enough to withstand these high pressures are not difficult to make with little more than ordinary care.

Table 6-3. Approximate Length of Usable Pipe Thread

Pipe	Distance Between Centers	Fittings	Center to Face of Fitting (Subtract)	Allowance for Threads (Add)	Overall Pipe Length
A	20"	2	$1\frac{5}{16}$"	$\frac{1}{2}$" + $\frac{1}{2}$"	$19\frac{11}{16}$"
B	43"	2-3	$1\frac{5}{16}$" + $1\frac{5}{16}$"	$\frac{1}{2}$" + $\frac{1}{2}$"	$41\frac{3}{8}$"
C	6"	3-4	$1\frac{5}{16}$" + $1\frac{5}{8}$"	$\frac{1}{2}$" + $\frac{1}{2}$"	$4\frac{1}{16}$"
D	24"	4-6	$1\frac{5}{8}$" + $1\frac{1}{16}$"	$\frac{1}{2}$" + $\frac{1}{2}$"	$22\frac{5}{16}$"
E	16"	5-6	$1\frac{1}{16}$" + $1\frac{5}{16}$"	$\frac{1}{2}$" + $\frac{1}{2}$"	$14\frac{5}{8}$"

The secret of making tight joints may be summed up as follows:

1. The threads must be clean.
2. Good-quality pipe-joint compound or Teflon tape must be used.
3. The joint must not be tightened so rapidly as to appreciably change the temperature of the metal.

Q. Are especially long threads favorable for tight joints?
A. No. The longer the thread, the greater the friction will be in the tightening process.

Q. Are perfect threads necessary to make tight joints?
A. No. A major manufacturer of pipe and fittings ran a series of experiments involving defective threads. It was found that with threads that were deliberately damaged to an extent 100 times greater than that for which most pipe is rejected, tight joints could be made without difficulty. These experiments proved that much of the pipe rejected because of defective threads could actually be used without impairing the serviceability of the piping system.

PIPE FITTING EXAMPLES

For most plumbing or pipefitting installations, a sketch should be made showing all the fittings and lengths of pipe necessary to complete the job. This sketch need not be elaborate, but simply a free hand pencil drawing. Fig. 6-14 shows a typical sketch for part of a boiler installation.

In order to avoid frequent changes of dies, it is best to make all lines of one size, when possible, before making up lines having a different size of pipe. In the installation in Fig. 6-14, the lines are all 3/4-in. pipe. The line in Fig. 6-14 consists of pipes *A, B, C, D,* and *E,* the pump connection *1,* and fittings *2, 3, 4, 5,* and *6.* Determine the overall dimensions of *A, B, C, D,* and *E* by preparing a table, using the measurements from the free-hand sketch and the dimensions given for standard fittings in Table 6-4. Your table should look somewhat like the following.

To determine the overall length of pipe *A*, notice that the distance between the face of the pump connection and the center of elbow 2 is 20 inches. This is shown in detail in Fig. 6-15. From Table 6-4, the distance from the center of a standard 3/4-in. elbow to its face is 15/16 in. From Table 6-3, the length of the thread that will be screwed into the fitting is 1/2 in. for a 3/4-in. pipe. Therefore, the total length of pipe *A* will be

$$20 - 1\tfrac{5}{16} + \tfrac{1}{2} + \tfrac{1}{2} = 19\tfrac{11}{16} \text{ inches}$$

Table 6-4. Dimensions of Standard Malleable-Iron Fittings

Size in.	1/8	1/4	3/8	1/2	3/4	1	1 1/4	1 1/2	2	2 1/2	3	3 1/2	4	4 1/2	5	6
A. in.	11/16	13/16	15/16	1 1/8	1 5/16	1 7/16	1 3/4	1 15/16	2 1/4	2 11/16	3 1/8	3 7/16	3 3/4	4 1/16	4 7/16	5 1/8
B. in.		3/4	13/16	7/8	1	1 1/8	1 5/16	1 7/16	1 11/16	1 15/16	2 3/16	2 3/8	2 5/8	2 13/16	3 1/16	3 7/16
C. in.			2 1/8	2 1/2	2 7/8	3 7/16	4 1/16	4 1/2	5 7/16	6 1/4	7 1/4		8 7/8			
D. in.			1 7/16	1 11/16	2	2 7/16	2 15/16	3 5/16	4 1/16	4 11/16	5 9/16		6 15/16			
E. in.		1	1 1/8	1 1/4	1 7/16	1 11/16	2 1/16	2 5/16	2 13/16	3 1/4	3 11/16	4	4 3/8			
F. in.	17/32	5/8	3/4	7/8	1 1/16	1 3/16	1 1/4	1 5/16	1 7/16	1 5/8	1 3/4	1 15/16	2		2 5/16	2 9/16
G. in.		1 1/16	1 3/16	1 5/16	1 1/2	1 11/16	1 15/16	2 1/8	2 1/2	2 7/8	3 3/16					
H. in.	1 1/8	1 5/16	1 7/16	1 5/8	1 7/8	2 1/8	2 1/2	2 11/16	3 3/16	3 13/16	4 1/2		5 11/16			
K. in.		15/16	1 1/16	1 3/16	1 5/16	1 1/2	1 11/16	1 7/8	2 1/4		3		3 3/4			
L. in.		5/8	11/16	13/16	15/16	1 1/16	1 1/4	1 3/8	1 11/16		2 1/8		2 1/2			

The lengths for pipes *B*, *C*, *D*, and *E* are determined in a similar manner, the only difference being that the measurements for fittings *4* and *5* are not listed in Table 6-4. These measurements, which vary according to the manufacturer, must be taken on the actual valve and union to be installed.

Fig. 6-16 shows the results of poor and good workmanship in the final makeup where the line is joined with a union. If pipe *E* is made too short, a gap at the union will make it difficult to bring the make-up joint together. Even if brought together, the system will be under an undue strain. The proper dimensions will result in the makeup joint springing into position snugly, with no appreciable stress or strain on any part of the system.

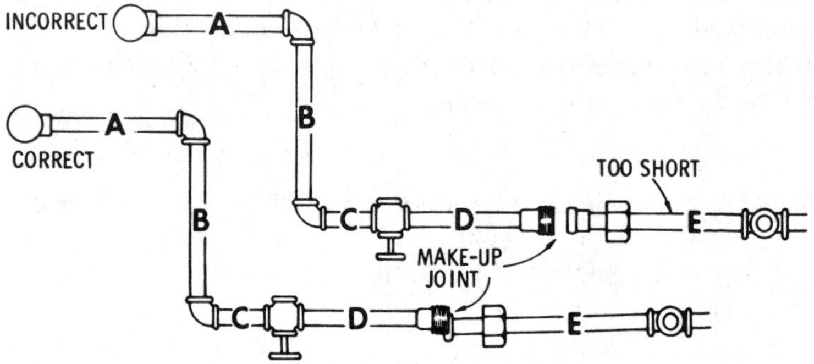

Fig. 6-16. The correct and the incorrect makeup of the feedline shown in Fig.6-14.

PIPE SUPPORTS

All piping should be supported in accordance with approved standards. Horizontal runs may be supported by hangers fastened to the ceiling or by wall brackets. The supports should be placed at frequent intervals—in no case more than 12 ft. apart. It is more desirable to support the pipe along concrete or masonry walls rather than wood floor joints since the wood joints may allow and amplify vibrations in the lines.

Vertical pipe runs are more difficult to support. A shoulder clamp bearing on the floor slab through which the pipes pass may be used. An alternate method is to build a supporting platform under the lower elbows to bear the weight of the risers. The upper horizontal runs should be well supported so as not to add any additional weight to the vertical runs. In addition, the vertical pipes should be clamped to adjacent walls or columns to hold the pipes rigid. These clamps should not be expected to bear any weight.

PIPE EXPANSION

Care must be taken in the design and installation of steam or hot-water pipes to provide for variations in length and form due to temperature changes. If these variations are not adequately taken care of, the system will be subject to undue stress and strain, resulting in possible damage to valves, joints, and fittings. The pipes should be securely anchored at certain points, while at other points, sliding or flexible hangers must be used. Expansion and contraction of the pipes is taken care of by such a method of support and by the addition of large-radius bends or expansion joints. Some of the commercially available pipe bends are shown in Fig. 6-17, while their critical dimensions are listed in Table 6-5.

CORROSION

In the treatment or prevention of corrosion, special consideration must be given to local conditions. In all cases where iron or steel pipes are exposed to moist air, they should be protected by waterproof and durable coatings. Internal corrosion is caused by the solvent or oxidizing properties of water and accelerated by the salts and gases (including air) dissolved in it. This makes the purification and treatment of boiler feed water necessary. The safest plan in this case is to consult a competent chemist experienced in the analysis and treatment of boiler feed water, and follow his recommendations.

PIPE SIZES

To efficiently convey water or steam through a pipe under pressure, the pipe must not be too small or there will be an undue

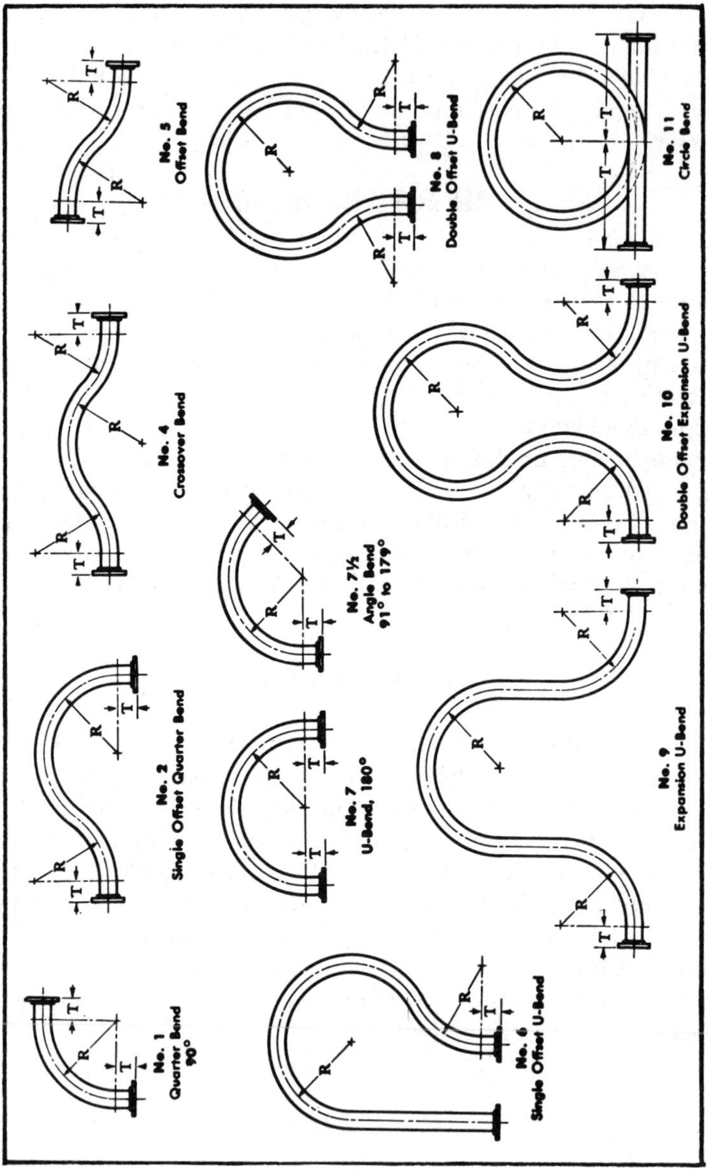

Fig. 6-17. Factory-made pipe bends.

Table 6-5. Radii and Tangents of Standard Pipe Bends

Column 1 Weight of Thickness of Pipe	Column 2 Size of Pipe	Column 3* Minimum Recommended Radius		Column 4* Minimum Tangent Including Pull Length		Column 5* Minimum Tangent After Cutting for Cranelap Joints		Column 6 Minimum Tangent After Cutting for Welding Ends Threaded Ends, Screwed Flanges	
	Inches	Cold Bend	Hot Bend	Cold Bend	Hot Bend	Cold Bend	Hot Bend	Cold Bend	Hot Bend
Standard or Heavier Weight	¼	1¼	1¼	3	3	. . .	. . .	1	1
	⅜	1⅞	1⅞	3	3	. . .	. . .	1¼	1¼
	½	2½	2½	3	3	2	2	1½	1½
	¾	3¾	3¾	3	3	2	2	1¾	1¾
	1	5	5	3	3	2	2	2	2
	1¼	6¼	6¼	3	3	2½	2½	2	2
	1½	7½	7½	3	3	3	3	2½	2½
	2	10	10	3	3	4	4	3	3
Standard or Extra Strong Weight for Cold Bend Process and Standard or Heavier Weight for Hot Bend Process	2½	12½	12½	18	12	5	5	4	4
	3	15	15	18	12	6	6	4	4
	3½	17½	17½	18	12	6	6	5	5
	4	20	20	18	12	6	6	5	5
	5	30	25	18	12	7	7	6	6
	6	42	30	18	12	7	7	7	7
	8	. . .	40	. . .	18	. . .	9	. . .	9
	10	. . .	50	. . .	18	. . .	12	. . .	12
	12	. . .	60	. . .	24	. . .	14	. . .	14
	14 OD	. . .	70	. . .	24	. . .	16	. . .	16
⅜-inch Thick†	16 OD	. . .	96	. . .	30	. . .	18	. . .	18
	18 OD	. . .	108	. . .	30	. . .	18	. . .	18
½-inch or Thicker	16 OD	. . .	80	. . .	30	. . .	18	. . .	18
	18 OD	. . .	90	. . .	30	. . .	18	. . .	18
	†20 OD	. . .	100	. . .	36	. . .	20	. . .	18
	†24 OD	. . .	120	. . .	36	. . .	24	. . .	18

Courtesy Crane Co.

*__Column 3__ radii preferably should equal or exceed dimensions shown; increased flattening and wall thinning usually result from smaller radii. **Column 4** dimensions are intended to cover bends that will be further fitted in the field. In **Column 5**, lapped stub ends welded on are recommended for sizes 2 inch and smaller, and for 18-, 20-, and 24-in. sizes.

†The table does not include sizes 20- and 24-in. of ⅜-in. wall thickness. Bending such pipe is not practical since "buckling" and excessive "wrinkles" will develop in the crotch section.

drop in pressure, resulting in insufficient flow. If the pipe is too large, the initial cost is impractical, and in the case of steam, an increase of condensation in the pipe renders the system inefficient. Table 6-6 lists the quantity of water capable of being delivered by various sizes of pipes under different pressures and conditions.

COPPER TUBING

Copper tubing and brass fittings have gained such an increasing use that they are now standard in many areas of the country. They offer several important advantages over the threaded-pipe systems. Some of these advantages are:

1. Flexible copper tubing can be easily bent and worked around obstructions. It can be run between studding and over and under electrical conduit and other pipes in much the same way an electrician installs an electric cable. It is usually unnecessary to tear holes in expensive finished walls and ceilings.
2. The longer lengths of flexible copper tubing eliminates many fittings. This type of tubing comes in coils of up to 250 ft. making continuous runs of that length possible without fittings.
3. The joints for copper tubing are easier and quicker to make than threaded joints. This can represent a considerable saving in time and labor.
4. Copper tubing is free from rust and much of the corrosion that causes steel pipe to become useless.
5. Flexible copper tubing can withstand freezing without bursting to a much greater degree than steel pipe.

Copper tubing is available in two major forms—flexible and rigid.

Flexible Tubing

Flexible copper tubing is available in coils ranging from 15 to 250 ft. in length. It is also manufactured in two types—Type "L" for indoor use and Type "K" for outdoor and underground installation.

Table 6-6. Water Flow in Steel Pipe

Condition of Discharge	Pressure in Main (psi)	Discharge, or quantity capable of being delivered (in cubic feet per minute) from the pipe under the conditions specified in the first column — Nominal Diameter (inches)							
		½	¾	1	1½	2	3	4	6
Through 35 feet of service-pipe—no back pressure	30	1.10	3.01	6.13	16.58	33.34	88.16	173.85	444.63
	40	1.27	3.48	7.08	19.14	38.50	101.80	200.75	513.42
	50	1.42	3.89	7.92	21.40	43.04	113.82	224.44	574.02
	60	1.56	4.26	8.67	23.44	47.15	124.68	245.87	628.81
	75	1.74	4.77	9.70	26.21	52.71	139.39	274.89	703.03
	100	2.01	5.50	11.20	30.27	60.87	160.96	317.41	811.79
	130	2.29	6.28	12.77	34.51	69.40	183.52	361.91	925.58
Through 100 feet of service-pipe—no back pressure	30	0.66	1.84	3.78	10.40	21.30	58.19	118.13	317.23
	40	0.77	2.12	4.36	12.01	24.59	67.19	136.41	366.30
	50	0.86	2.37	4.88	13.43	27.50	75.13	152.51	409.54
	60	0.94	2.60	5.34	14.71	30.12	82.30	167.06	448.63
	75	1.05	2.91	5.97	16.45	33.68	92.01	186.78	501.58
	100	1.22	3.36	6.90	18.99	38.89	106.24	215.68	579.18
	130	1.39	3.83	7.86	21.66	44.34	121.14	245.91	660.36
Through 100 feet of service-pipe and 15 feet vertical rise	30	0.55	1.52	3.11	8.57	17.55	47.90	97.17	260.56
	40	0.66	1.81	3.72	10.24	20.95	57.20	116.01	311.09
	50	0.75	2.06	4.24	11.67	23.87	65.18	132.20	354.49
	60	0.83	2.29	4.70	12.94	26.48	72.28	146.61	393.13
	75	0.94	2.59	5.32	14.64	29.96	81.79	165.90	444.85
	100	1.10	3.02	6.21	17.10	35.00	95.55	193.82	519.72
	130	1.26	3.48	7.14	19.66	40.23	109.82	222.75	597.31
Through 100 feet of service-pipe and 30 feet vertical rise	30	0.44	1.22	2.50	6.80	14.11	38.63	78.54	211.54
	40	0.55	1.53	3.15	8.68	17.79	48.68	98.98	266.59
	50	0.65	1.79	3.69	10.16	20.82	56.98	115.87	312.08
	60	0.73	2.02	4.15	11.45	23.47	64.22	130.59	351.73
	75	0.84	2.32	4.77	13.15	26.95	73.76	149.99	403.98
	100	1.00	2.75	5.65	15.58	31.93	87.38	177.67	478.55
	130	1.15	3.19	6.55	18.07	37.02	101.33	206.04	554.96

Either flare- or solder-type fittings can be used on flexible copper tubing. Flare fittings are generally more expensive than the comparable solder type, but have the advantage of being installed without the use of heat. This is a decided advantage in old construction, where the use of a torch might be a definite fire hazard. For new construction, and in areas where the use of a soldering torch is not dangerous, solder fittings are recommended. In fact, the plumbing code in certain communities prohibits the use of flare fittings. Always consult the proper authorities to determine the code in effect in that area.

Copper tubing is easily cut with either a hacksaw or a tubing cutter, but care must be taken to make the cut square. This is especially important if a flared fitting is to be installed. A square cut will permit a perfect flare to be formed, whereas a crooked cut will almost certainly result in a poor flare that will leak. After the tubing is cut, the burr should be removed from the inner diameter. Most tubing cutters have a reamer for this purpose. A separate hand reamer can be used when the tubing has been cut with a hacksaw. Fig. 6-18 shows a typical tubing cutter with attached reamer.

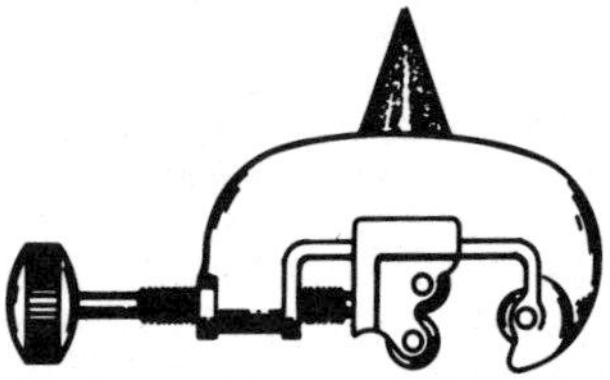

Fig. 6-18. A copper tube cutter.

Flexible tubing can be bent either by hand (if care is taken), by using bending springs (Fig. 6-19), or by means of a tube bender (Fig. 6-20). The use of bending springs, either internal or external, permits a much sharper bend to be made without collapsing the tubing. A tube bender also permits a sharper bend as well as a more even curve than is possible when bending by hand. When a very sharp bend must be made, an elbow, such as the flared unit shown in Fig. 6-21, should be used. A solder-type elbow could also be used here if conditions permit the application of heat from a soldering torch.

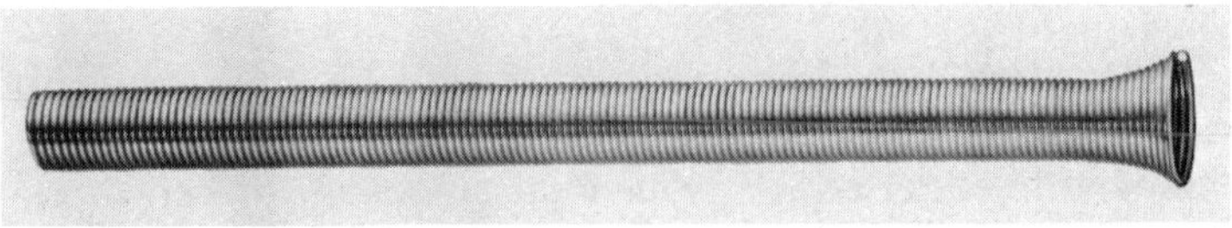

Fig. 6-19. A spring-type tube bender.

Reducing fittings are available when it is desirable to change from one size of tubing to another. Fig. 6-22 shows the change from ¾ in. to ½ in. through the use of a reducing coupling. A

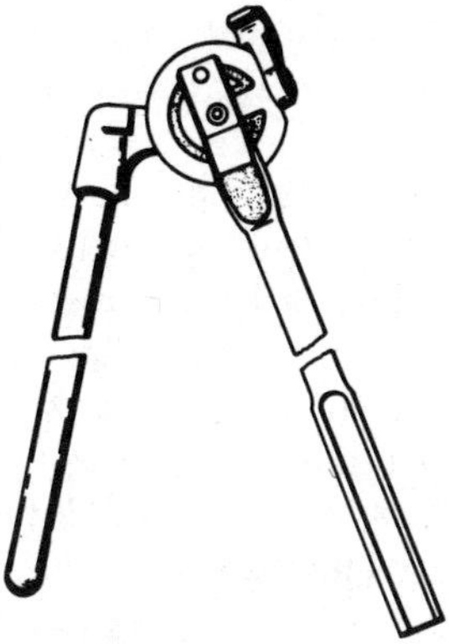

Fig. 6-20. A lever-type copper tube-bending tool.

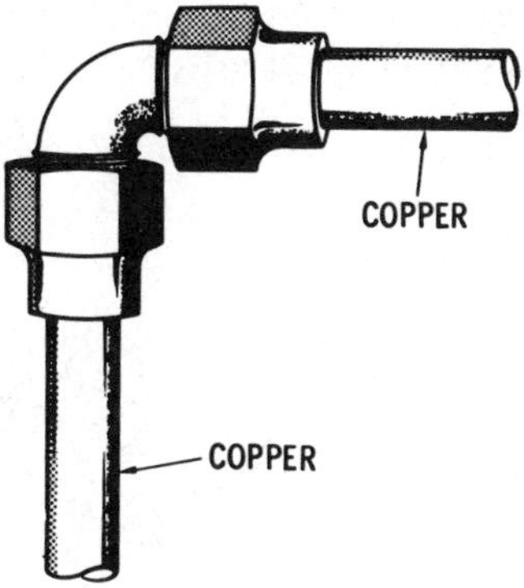

Fig. 6-21. A flare-type copper tube elbow.

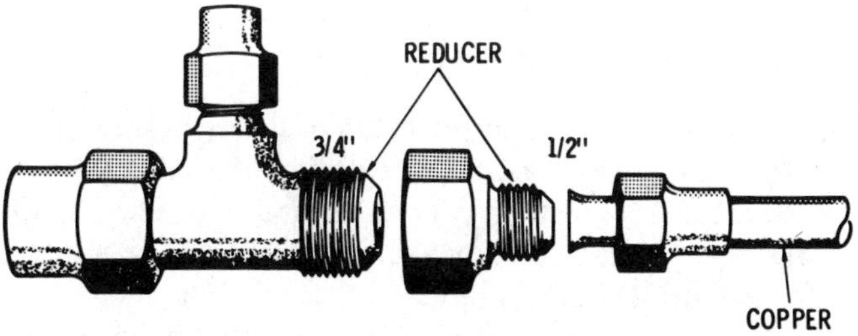

Fig. 6-22. A flare-type copper tube-reducing fitting.

reducing T can be used in this situation. Reducing elbows, coupling, T's, Y's, etc., are available in practically any combination and size of reduction that will be needed.

Fittings are also available with one or more of the openings threaded for standard iron pipe and the remaining openings arranged for connection to copper tubing. In addition, adapters are available to screw in or on standard pipe fittings to accept copper tubing. The variety, sizes, and combinations possible will take care of nearly any conceivable installation problem.

The flare on copper tubing is made with a flaring tool such as the one in Fig. 6-23. This consists of two bars held together by a wing nut and bolt; the bars are provided with holes for the various sizes of tubing. A yoke containing the forming die is slipped over the bars, and the handle is turned to produce a flare. This tool is widely used because of its simplicity an its ease of performing the flaring operation, and because the flares it makes are uniformly excellent in producing a tight seal. Before applying pressure with the forming tool, place a drop of oil on the end of the tube.

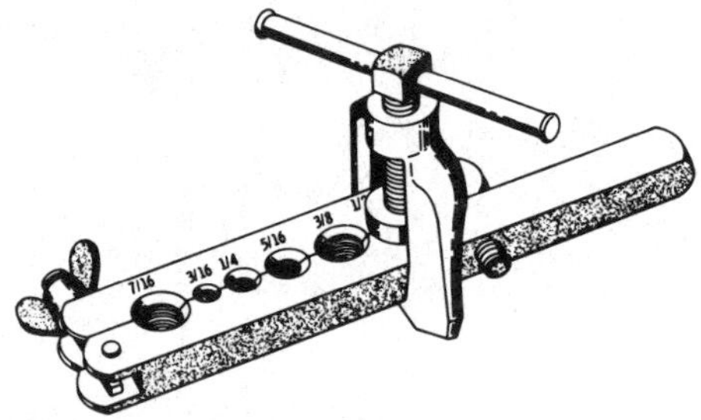

Fig. 6-23. A flaring tool.

Rigid Tubing

Rigid copper tubing is used extensively in plumbing work, especially in new construction. A full line of solder fittings and valves is available for installing this type of tubing. Three grades or types are manufactured—Type L and Type M for indoor and Type K for outdoor and underground installation.

Bending rigid tubing should not be attempted on the job unless a power bender is available. Even with power equipment, the minimum radius of the bend is relatively large compared to the bend possible with flexible tubing. In most cases, fittings are used to change the direction of straight sections of tubing.

Fittings are not difficult to install, but care must be taken to make sure a leakproof joint has been made. Before assembly, the inside surface of the fitting which will form the solder joint, and the external surface of the end of the tubing, should be made bright with steel wool or emery cloth. These bright surfaces should then be coated with soldering flux. After the tubing is slipped into the fitting, heat is applied to the area by a torch such as a blowtorch

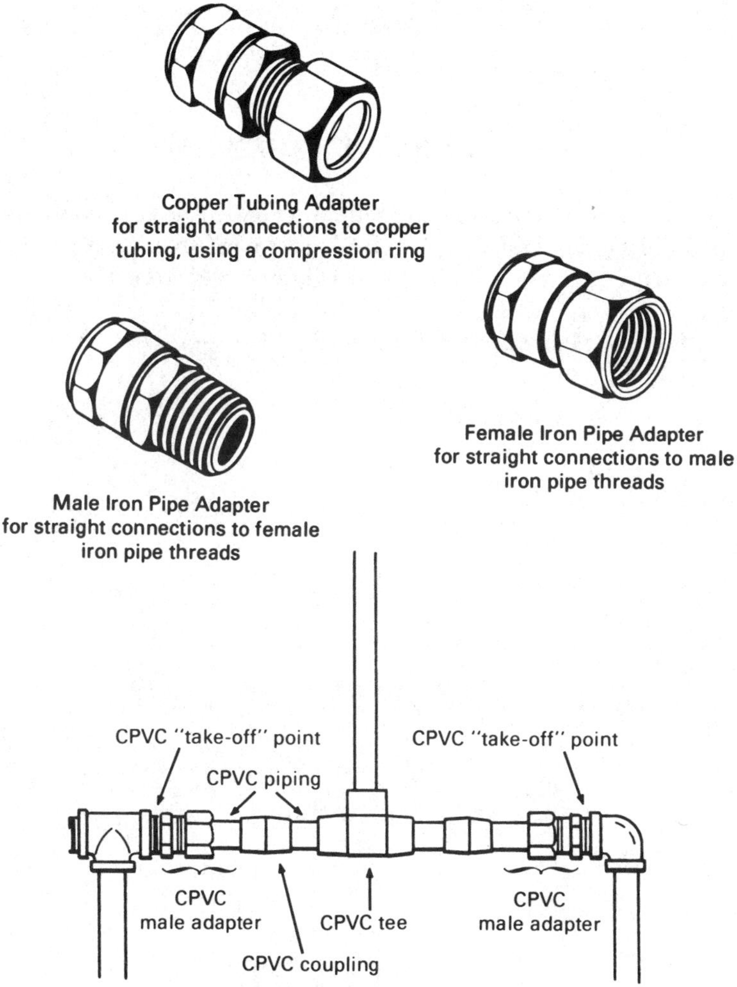

Fig. 6-24. Typical transition fittings and their uses.

or propane torch. When the metal becomes hot enough, wire solder is applied to the rim of the fitting. Capillary attraction will pull the liquified solder up into the space between the fitting and the tubing, and it will distribute itself evenly. Practice will indicate when a leak-proof, strong joint has been made. This is generally true when a complete and smooth fillet of solder forms between the rim of the fitting and the surface of the tube. An excess of solder is unnecessary and is to be avoided. A relatively small amount of solder is all that is needed to make a perfect joint.

PLASTIC PIPE

The use of plastic pipe, as mentioned earlier, is now common, particularly for drainage lines for highly corrosive waste. Plastic pipe is used for complete water distribution systems in homes and industries, especially for underground lawn-sprinkling piping. Fittings—called transition fittings (Fig. 6-24)—are also available for joining.

CHAPTER 7

Roughing-in

Roughing-in as applied to plumbing and pipe fitting is a term used for the installation of concealed piping or fittings at the time a building is under construction or being remodeled. As the building nears completion, the final connections of plumbing or heating fixtures are made.

Roughing-in is very important. If the piping is not roughed-in correctly, the final connections may be very difficult. Plans or blueprints show room dimensions and equipment locations, but they do not show at what elevations or locations piping must be stubbed through a wall for final connections.

Roughing-in drawings are furnished by equipment suppliers, and these drawings show not only the correct location of the roughed in piping but also show where to install any necessary backing boards behind the finished walls. A wall-hung lavatory hangs on a hanger or bracket, and this hanger in turn must be fastened to the wall using long wood screws that will go through the plaster or drywall and into a backing board.

The plans for a residence may show a bathroom with fixture locations, but usually no definite measurements or locations are given. The location of the stub wall at the end of the bathtub is determined by the length of the tub. This type of bathtub is called a *recessed tub* because it is placed against the stud walls, and the lath and plaster or drywall is then installed.

Fig. 7-1 shows an *architectural* plan of a typical bathroom. Plans and blueprints use the symbol ′ to indicate feet and ″ to indicate inches. The plan in Fig. 7-1 shows the bathroom to be 6′-0″ wide by 8′-0″ long; both are finished room measurements. Fig. 7-2 shows a *mechanical* plan of the same bathroom; here again, the fixture locations are shown, water and waste piping is indicated, but no measurements are given. The exact location of the water closet and the lavatory must be determined by the plumber. The minimum space to be allowed for the installation of a water closet is generally conceded to be 32 in. If there is sufficient space, 36 in. is preferable.

As shown in Fig. 7-1 and Fig. 7-2 there is ample room for a 36-in. space for the water closet. At the time the plumber is roughing-in the building, only the stud walls are up, so he must determine where the finished walls will be. If the walls are to be a typical drywall and ceramic tile, for instance, he would add ⅝ in. for drywall and ⅜ in. for ceramic tile and locate the center of the waste piping for the water closet 19 in. from the outside stud wall (18″ + 1″).

In order to find the distance out from the 8 in. wall to the center of the same waste piping, the plumber would then check the roughing-in measurements for the water closet. A typical rough-in sheet for a water closet is shown in Fig. 7-3. This shows the center of the waste piping to be 12 in. from the finished wall or 13 in. from the rough wall (12″ + ⅝″ + ⅜″). Using these methods the center of the waste piping from both walls has now been determined.

ROUGH-IN DRAWING

Water-closet rough-in sheets are similar to Fig. 7-3. All the necessary information for roughing-in a water closet is shown on the roughing-in sheet furnished by the manufacturer. Rough-in

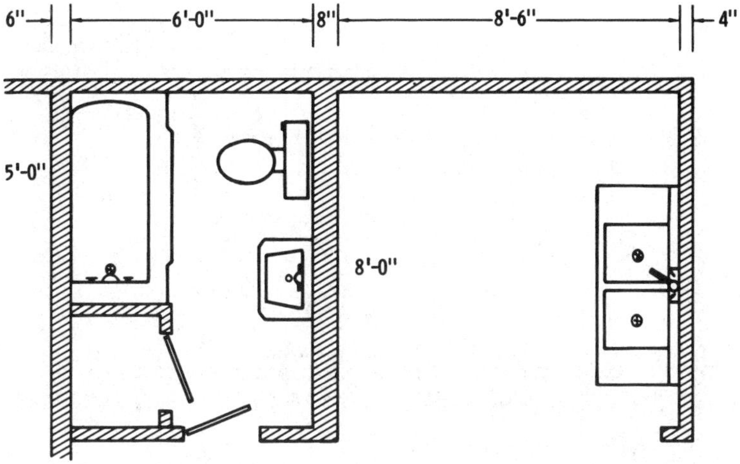

Fig. 7-1. A partial architectural plan of a building showing the kitchen and bathroom areas.

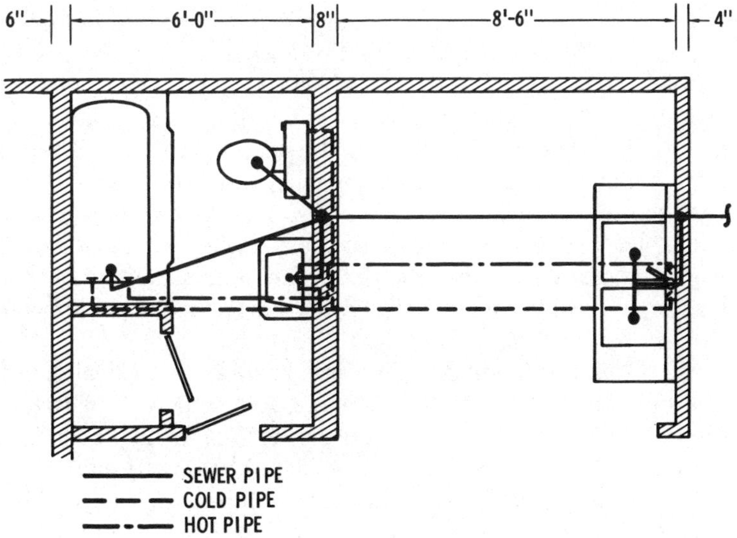

Fig. 7-2. A partial mechanical plan of a building showing the kitchen and bathroom areas.

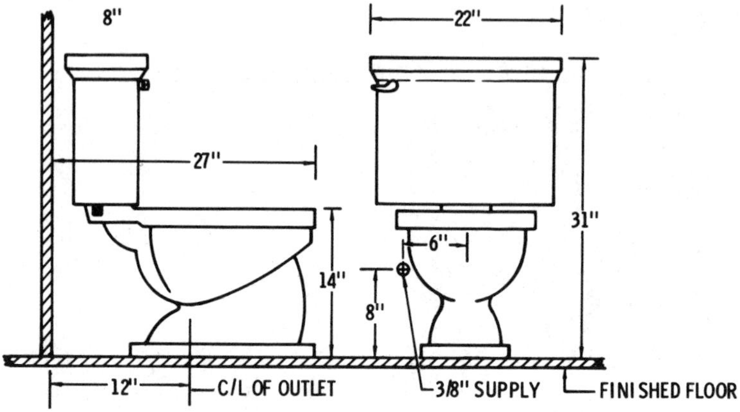

Fig. 7-3. Water closet rough-in sheet.

sheets vary, depending on the manufacturer and type of fixture, but they will show the important points as outlined in the preceding section.

The center of the lavatory waste piping may be determined in any one of several ways. A medicine cabinet may already be framed in, in which case it would be necessary to center the lavatory on the medicine cabinet for the finished appearance to look right. Another way is to scale the plan. If a rule is laid on the plan, it shows the distance from the outside wall to be 4′-6″ to the center of the lavatory. A job condition can sometimes make scaling a plan impractical. In this instance scaling would probably work out very well, due to the size of the room. The center of the lavatory waste piping would be 55 in. from the outside *rough* wall (54″ + ⅝″ + ⅜″).

Fig. 7-4 is a rough-in drawing for a lavatory that is wall hung. The drawing for this particular make, size, and type of fixture shows the center of the waste piping to be 17 in. from the finished floor. If the plumber is working from a rough floor, he must allow for the thickness of the material which will be added to produce the finished floor. Assuming this to be 2 in. (15/8″ grout + ⅜″ ceramic tile), the center of the waste piping would be 19 in. above the *rough* floor.

The center of the waste piping for the bathtub is determined when the waste and overflow fitting is installed on the tub. The

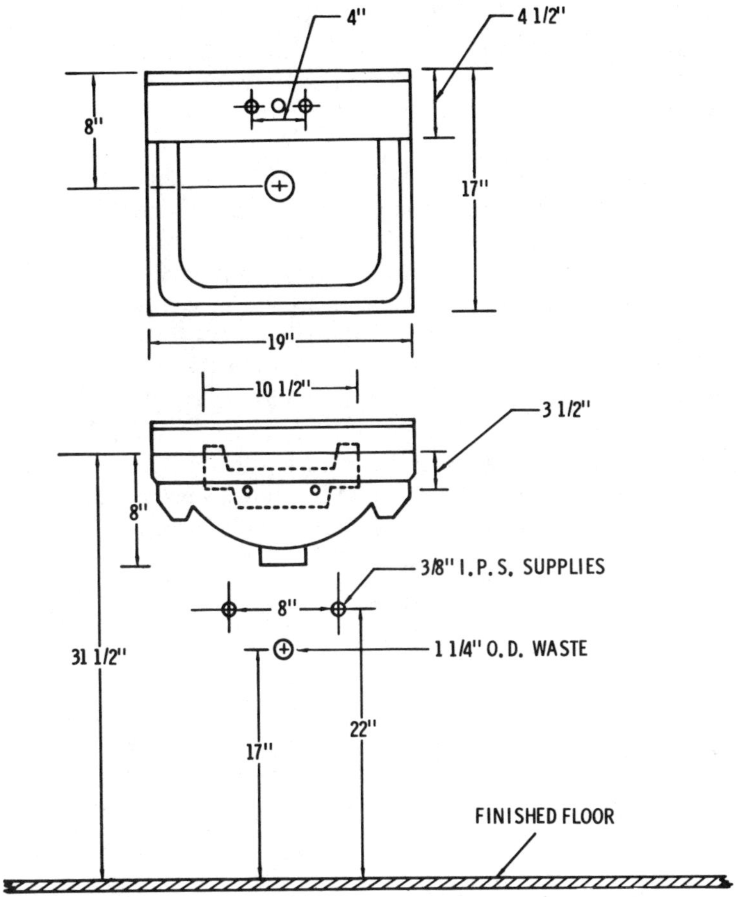

Fig. 7-4. Lavatory rough-in sheet.

drain piping, with a trap, is then extended to this point and connected to the waste and overflow fitting.

Rough-in drawings show the location of the piping needed for final connection to the fixtures. The water-supply location for the water closet is shown in Fig. 7-4 as being 6 in. to left of center of the fixture and 8⅞ in. above finished floor. The mechanical plan of the piping shown in Fig. 7-2 indicates that the water piping to the water closet shall be ½ in.; however, the final connection will be ⅜ in. Common practice is to use a reducing fitting (½″ × ⅜″) behind the finished wall and stub through the wall with ⅜ in. pipe.

The rough-in drawing, Fig. 7-4, shows the water connections to the lavatory to be 8 in. on center (4 in. to each side of center) and 3/8 in. in size. Here again, it is common practice to use a reducing fitting behind the finished wall and stub through the wall with 3/8 in. pipe.

The rough-in drawing for the bathtub, Fig. 7-5, shows the water piping to be 1/2 in., 8 in. center to center, and centered on the tub waste and overflow openings. The tub valves, minus the finish chrome plated parts (trim) and the shower piping, are installed as part of the roughing-in operation. A pipe fitting with a 1/2 in. f.p.t. (female pipe thread) must be left behind the finished wall to receive the chrome plated shower arm and the finish trim are installed as part of the finish operation when the fixtures are set.

A typical roughing-in sheet for a bathtub shows that backing boards are usually necessary for securing the curtain rod to the wall, as well as the piping rough-in locations. The tub faucet, or mixing valve (called an over-rim filler), should be centered on the waste and overflow connection.

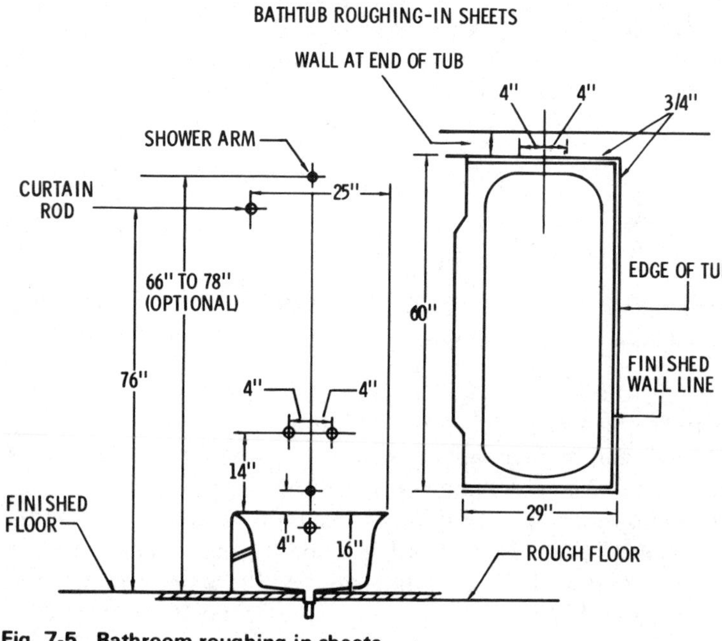

Fig. 7-5. Bathroom roughing-in sheets.

Recessed tubs are identified as L.H. (left hand) or R.H. (right hand). Facing the tub front, or apron, if the drain connection is at the left end of the tub, it is a L.H. tub. If the drain is at the right end, it is a R.H. tub.

Corner tubs Fig. 7-6 are identified in a different manner. Facing the front or apron of the tub, if the corner is on the left end, it is a L.H. corner tub; if the corner is on the right, it is a R.H. corner tub.

A typical kitchen sink rough-in sheet does not show water or waste opening rough-in dimensions. Job conditions plus a great variety in the selections of faucets and drain connections make it impractical to show the rough-in measurements. In almost all cases, the suggested measurements shown in Fig. 7-7 will enable connection to either a double sink waste or to a single waste and a garbage disposer. If the waste opening is roughed in at approximately the center of either of the sink compartments, final connection will be easier. The hot- and cold-water piping can be roughed-in through the wall at approximately 16 in. above the finished floor or roughed-in through the floor for final connection.

Courtesy Eljer Plumbingware

Fig. 7-6. A tub classified as right-handed.

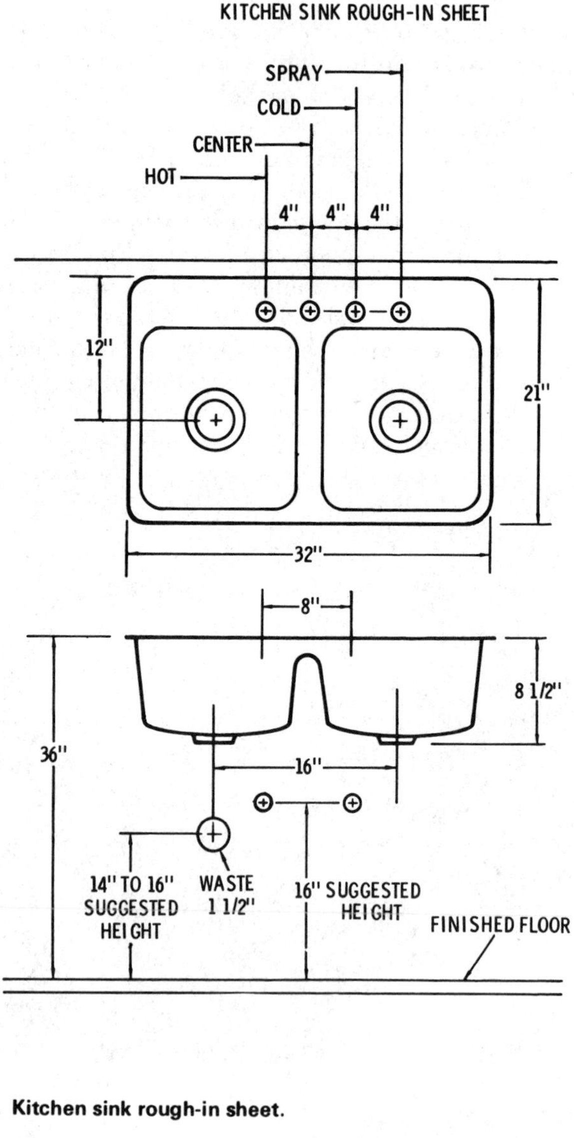

Fig. 7-7. Kitchen sink rough-in sheet.

ISOMETRIC DRAWING

The purpose of an isometric drawing is to give a view of all parts of a system. If you can imagine that you are standing above and slightly to one side of the bathroom we have been describing and that you are looking down on the finished rough-in piping, it would look similar to the view shown in Fig. 7-8. Isometric drawings are a valuable aid to estimating or ordering fittings since virtually every fitting used on the job will be shown in the drawing, if the drawing is done correctly. Fig. 7-8 is an isometric drawing showing which part of the soil, waste, and vent piping uses cast-iron soil pipe and which part uses D.W.V. (drainage, waste, and vent) copper piping.

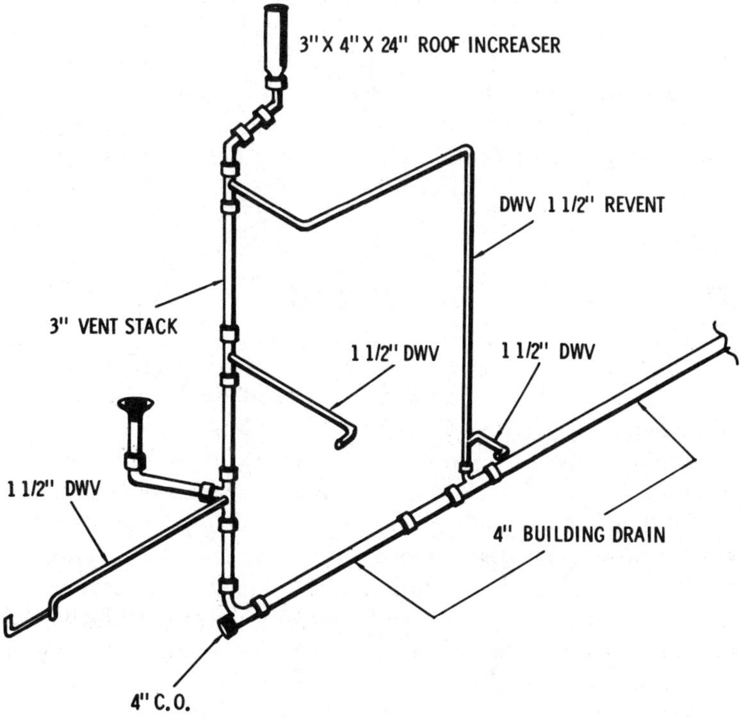

Fig. 7-8. A typical isometric drawing.

Roughing-in a building is not a big job—it is a series of little jobs. A ten-story building seems like a big job, but if one looks at it as ten one-story jobs, it doesn't seem so formidable.

Fig. 7-2 shows a partial mechanical plan of a bathroom and kitchen area of a residence. The experienced plumber can take the plan in Fig. 7-2, make a sketch similar to the one in Fig. 7-9, and order all the soil and waste piping needed.

S.H. (single hub) soil pipe:	5-ft. lengths
	10-ft. lengths
D.G. (double hub) soil pipe:	5-ft. lengths
	2½-ft. lengths (30″)

The 30-in. lengths provide short pieces with hubs with a minimum of waste. Because of space limitations, we can show only a partial plan of this building, but it is not difficult to see the ease of estimating and ordering material using the information on the plans and the isometric drawings. A study of the complete plans would show a 3-ft. crawlspace under the house, and the room ceiling heights are shown as 8 ft.

MATERIAL

An isometric drawing, such as the one shown in Fig. 7-9, makes it possible to determine the materials needed for the job. Building drain and waste vent stack materials would include cast-iron soil pipe and fittings. Two-inch and smaller waste and vent piping consists of D.W.V. weight copper pipe and fittings. The water piping is made up of Type L hard copper pipe and wrought copper fittings. The experienced plumber will notice immediately that the soil and waste stack should be reduced to 3 in. in order to fit all the piping into an 8-in. wall. Most city and state codes would permit a 3-in. vent through the roof, but since some codes would require a 4-in. extension (to keep the vent from frosting over in extreme cold weather), we will include a 4-in. vent through the roof. Using the information given in Fig. 7-9, a material list should be made such as the one below.

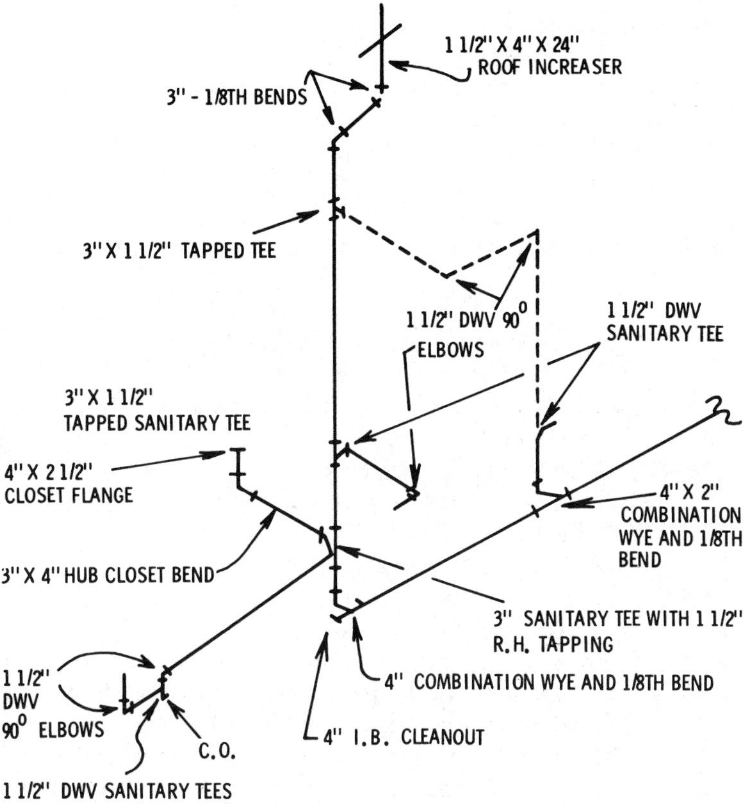

Fig. 7-9. An isometric drawing showing materials to be used.

Fixtures:

1—5-ft. L.H. (left-hand) recess tub with over-rim filler, shower, and trip lever drain.

1—12-in. c.c. (close-coupled) closet combination with white seat, supply and stop.

1—19 ″ × 17″ V.C. (vitreous china) lavatory complete with centerset faucet, pop-up drain, supplies, stops and C.P. “P” trap.

1—32″ × 21″ stainless steel sink complete with basket strainers, connected sink waste, trap, and single lever type faucet with spray.

Material:

10 ft. (1 length) 4-in. S.H. soil pipe.
5 ft. 4-in. D.H. soil pipe.
5 ft. 3-in. S.H. soil pipe.
10 ft. 3-in. D.H. soil pipe (5′ lengths).
2—30-in. lengths 3-in. D.H. soil pipe.

(It is always better to order a little extra soil pipe to allow for cracked hubs, etc., the extra material can always be returned to the shop.)

1—4″ × 2″ combination wye and 1/8 bend.
1—4″ × 3″ combination wye and 1/8 bend.
1—3 in. sanitary tee with 1 1/2-in. R.H. tapping.
1—3″ × 4″ hub closet bend.
1—4″ × 2 1/2″ (deep) closet flange.
2—3″ × 1 1/2″ tapped sanitary tees.
2—3 in. 1/8 bends.
1—3″ × 4″ × 24″ calk roof increaser.
1—4 in. I.B. (iron-body) cleanout.

The amount of lead is estimated in this matter: 1 lb. for each inch/joint. A 3-in. joint would need 3 lbs of lead. Oakum is estimated at 1 lb. of oakum for each 10 lbs of lead. In actual practice it will be found that the amount of lead used will be somewhat less than the 1 lb. per inch formula but this formula will allow for spillage, etc. In the partial plan we are working with we would have, according to our isometric drawing (Fig. 7-8).

7—4-in. lead joints	=	(7 × 4)	28 lbs. lead
12—3-in. lead joints	=	(12 × 3)	36 lbs. lead
1—2-in. lead joint	=	(1 × 2)	2 lbs. lead
		Total	66 lbs. lead

The amount of oakum needed, using the formula given above, would be 70 lbs. (rounding off the 66 lbs. of lead to the nearest tenth).

The waste and vent piping for the kitchen sink would require:

2—1½-in. DWV 90° elbows
1—1½-in. DWV M.I.P. (male iron pipe) copper adapter
1—1½-in. DWV copper sanitary tee
20 ft. (approx.)—1½-in. DWV copper pipe

The waste piping for the lavatory would require:

1—1½-in. DWV M.I.P. copper adapter
1—1½-in. DWV 90° elbows
3 ft. (approx.)—1½-in. DWV copper pipe

The waste piping for the bathtub would require:

1—1½-in. DWV M.I.P. adapter
2—1½-in. DWV copper 90° elbows
1—1½-in. DWV copper tee
1—1½-in. DWV copper fitting cleanout
6 ft. (approx.)—1½-in. DWV copper pipe

Approximately ½ lb. of 50/50 solder would be required for the DWV waste and vent piping. If the waste and vent piping were installed as shown in the drawings, the bathtub and water closet would be wet-vented by the lavatory, which would be permissable by virtually all local plumbing codes. A certain amount of knowledge will be called for when estimating the waste piping material needed. For instance, valves may not show on the plans at the bathtub and kitchen sink locations, but good plumbing practice (and most state and city codes) demands that valves be installed at these locations. Final connections at the bathtub and kitchen sink locations will require ½-in. openings; the waste closet and the lavatory will require ⅜-in. openings.

Water piping material for the four fixtures shown:

2—¾″ × ¾″ × ½″ copper to copper tees
1—¾″ × ½″ × ¾″ copper to copper tee
2—¾″ × ½″ × ½″ copper to copper tees
1—¾-in. copper to copper 90 elbow
4—½-in. copper sweat stops (valves)

3— ½-in. copper × ⅜ & F.I.P. (female iron pipe) elbows
2— ½-in. copper × ½ & FIP elbows
12 (approximately)— ½-in. copper to copper elbows
30 ft. (+ −)— ½-in. type-L hard copper pipe (tubing)
15 ft. (+ −)— ¾-in. type-L hard copper pipe (tubing)
½ lb.—50/50 solder

(Note that stops, or valves, are estimated for the water closet and lavatory with the fixtures.)

The above material can be ordered or estimated easily using the isometric drawing shown in Fig. 7-9.

Fig. 7-10 is an isometric drawing of the hot and cold water piping shown in Fig. 7-2.

HEATING SYSTEM ROUGH-IN

Fig. 7-11 shows the installation of a typical section of baseboard radiation used with a hot-water heating system. Roughing-in is important here in that the piping to supply the radiation and the

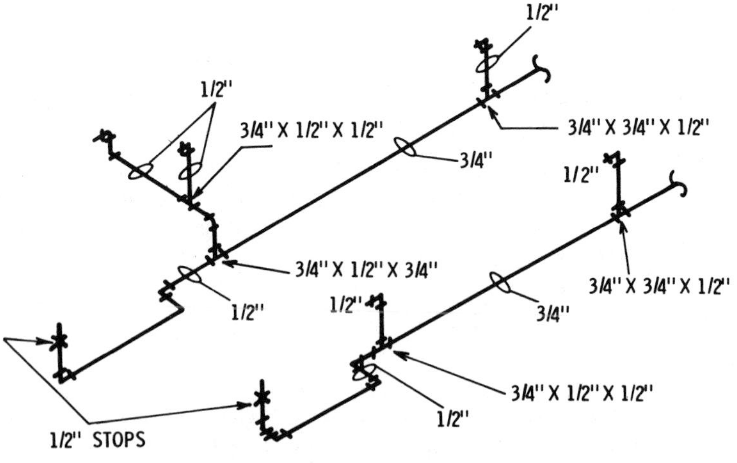

Fig. 7-10. An isometric drawing of the hot and cold water piping shown in Fig. 2.

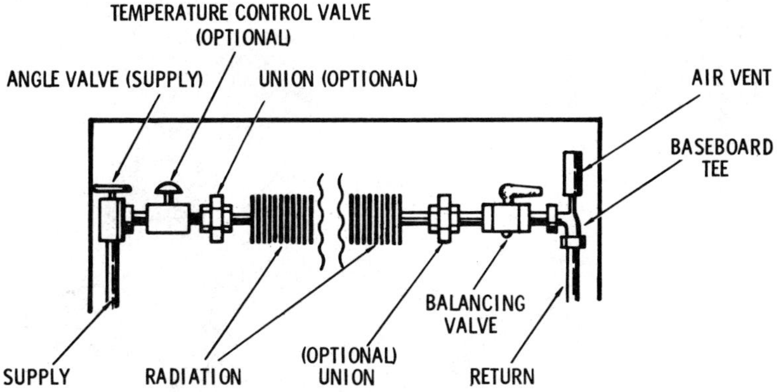

Fig. 7-11. A typical section of baseboard radiation used with a hot water heating system.

return piping must be brought through the wall or floor at a point where it will be concealed by the radiation cover. If valves or balancing cocks are used to control the radiation, sufficient room must be allowed to permit their installation. In general, the rough-in drawings furnished by the manufacturers of baseboard radiation, cabinet heaters, and convectors are intended only to show the installation of the equipment. They do not show the piping connections because the piping connections may vary due to the design of the system.

Some state and city plumbing codes prohibit a number of fixtures, as seen in Fig. 7-12. These include the full "S" traps, saddles, and bell trap floor drains. Trough-type urinals are also prohibited in many areas.

The fittings shown on the following pages of this chapter are presented to enable the reader to be able to obtain measurements of these fittings for layout work.

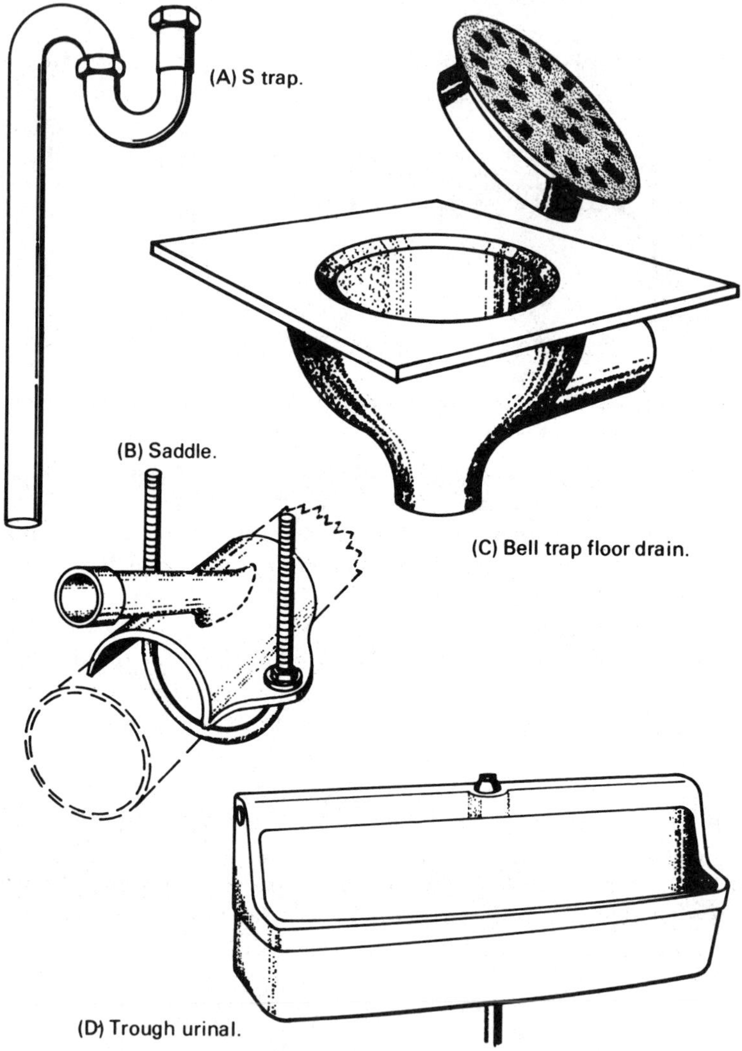

Fig. 7-12. Several plumbing items which are prohibited by some state and city plumbing codes; many municipalities accept these things.

Dimensions for 90° Elbows

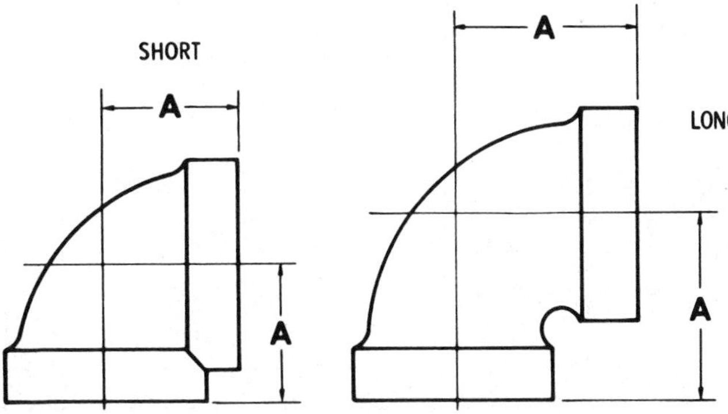

90° ELBOW, SHORT TURN		90° ELBOW, LONG TURN	
Size	A	Size	A
1¼	1⅞	1¼	2⅜
1½	2	1½	2 11/16
1½ × 1¼	2	2	3 7/16
2	2⅝	2½	3 11/16
2 × 1½	2⅛	3	5⅞
2½	2 13/16	4	6¾
3	3⅜	5	6⅛
4	3 15/16	6	8 7/16
5	4⅝	7	9½
6	5 11/32	8	1⅞
7	5 11/16	10	12¼
8	6½	12	13
10	7⅞		
12	9		

Dimensions for 45° Elbows

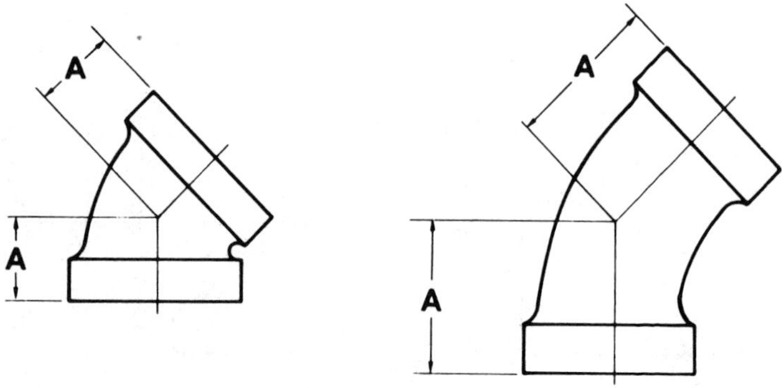

45° ELBOW, SHORT TURN		45° ELBOW, LONG TURN	
Size	A	Size	A
1 1/4	1 5/16	1 1/4	1 3/4
1 1/2	1 1/2	1 1/2	1 7/8
2	1 5/8	2	2 1/4
2 1/2	1 11/16	2 1/2	3
3	2 1/4	3	4 5/8
4	2 7/16	4	4 5/8
5	2 11/16	5	5 7/16
6	3	6	6
7	3 3/8	7	6 1/4
8	3 1/2	8	6 3/4
10	4 7/16	10	7 1/2
12	5 1/2	12	8 1/4

Dimensions for 22½° and 11¼° Elbow

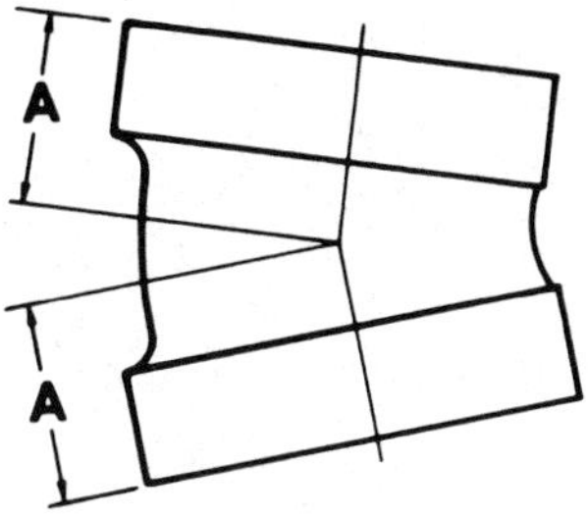

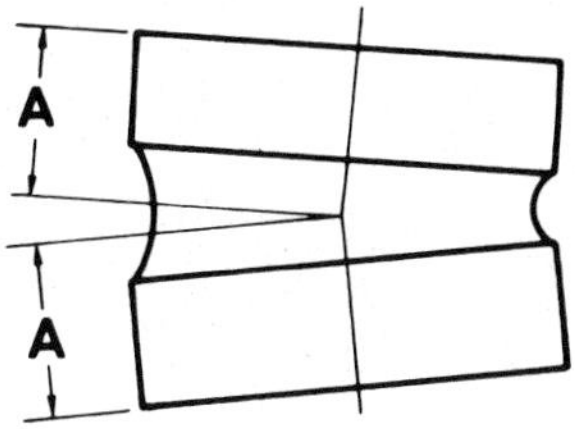

22½° ELBOW		11¼° ELBOW	
Size	A	Size	A
1¼	1 1/16	1¼	1 1/16
1½	1¼	1½	1⅛
2	1 7/16	2	1⅜
2½	1½	2½	1 7/16
3	1⅞	3	1⅝
4	1⅞	4	1 11/16
5	2	5	1⅞
6	2 3/16	6	2
7	2½	7	2 1/16
8	2⅞	8	2⅛
10	3¾	10	3
12	4		

Dimensions for 5⅝° and 60° Elbows

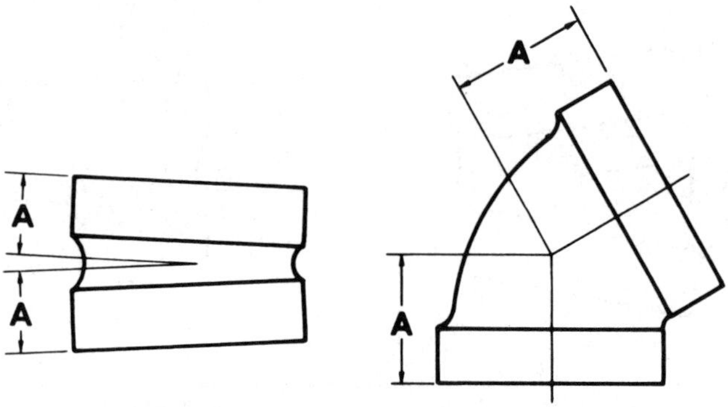

5⅝° ELBOW		60° ELBOW	
Size	A	Size	A
1¼	1 1/16	1¼	1½
1½	1⅛	1½	1½
2	1¼	2	1⅞
2½	1½	2½	2¼
3	1⅝	3	2½
4	1 11/16	4	2⅞
5	1¾	5	3¼
6	1⅞	6	3⅝
7	1 15/16	7	4
8	2	8	4¾
10	2¼	10	5¼

Dimensions for Special Elbows

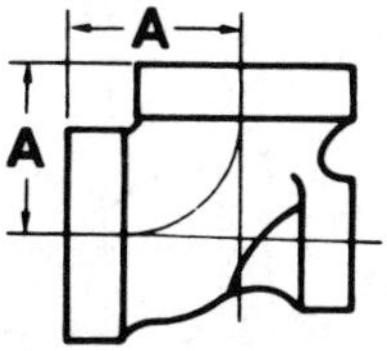

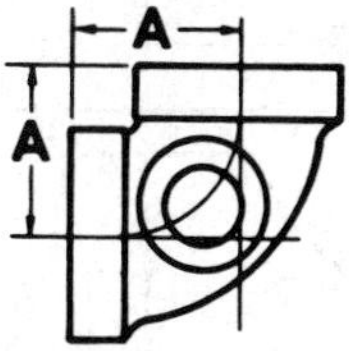

90° ELBOW WITH HEEL OUTLET		90° ELBOW WITH SIDE OUTLET	
Size	A	Size	A
4	3 15/16	4	3 15/16

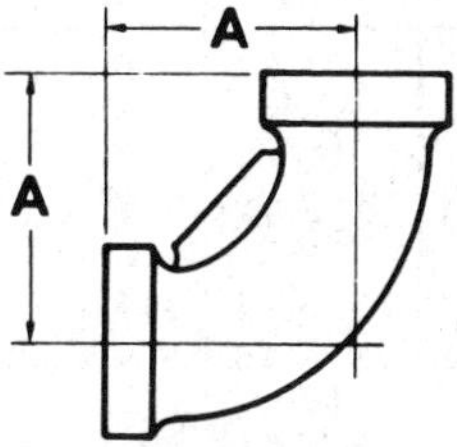

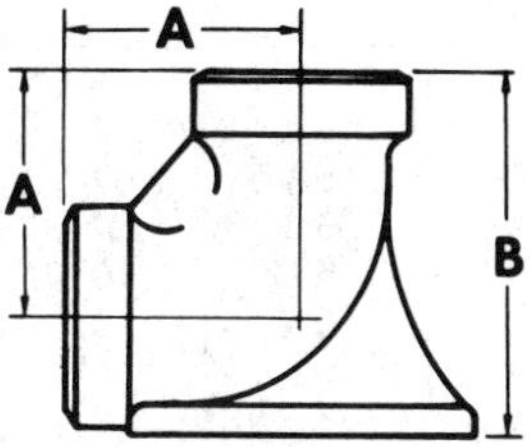

90° ELBOW WITH CLEANOUT		90° BASE ELBOW WITH CLEANOUT		
Size	A	Size	A	B
2	3 7/16	2	3 7/16	5 1/8
3	5 7/8	3	5 7/8	8 1/8
4	6 3/4	4	6 3/4	9 1/2
5	6 1/8	5	6 1/8	9 1/2
6	8 7/16	6	8 7/16	12 7/16

Dimensions for Special Elbows

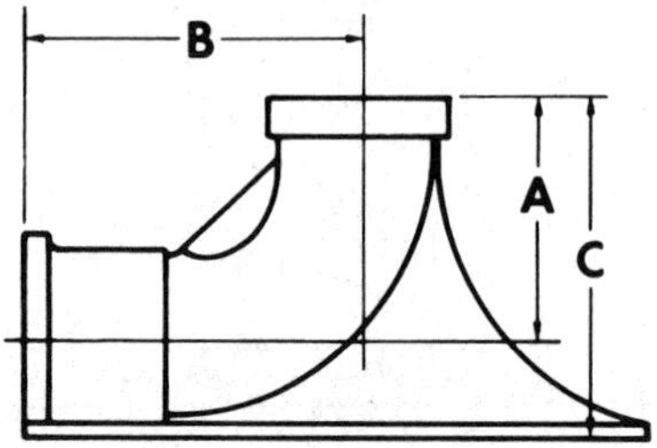

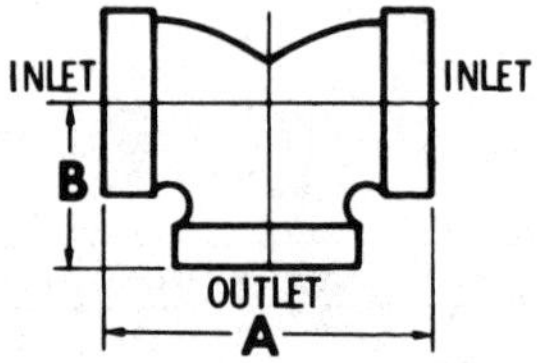

90° BASE ELBOW, WITH CLEANOUT AND SOIL-PIPE HUB CONNECTION				3-WAY ELBOW		
Size	A	B	C	Size	A	B
3 × 2	$3\frac{7}{16}$	$6\frac{3}{4}$	$6\frac{1}{2}$	$1\frac{1}{4}$	$4\frac{3}{4}$	$2\frac{3}{8}$
3	$5\frac{7}{8}$	7	$8\frac{1}{8}$	$1\frac{1}{2}$	$5\frac{1}{2}$	$2\frac{3}{4}$
4 × 3	$5\frac{7}{8}$	$7\frac{1}{2}$	$8\frac{5}{8}$	2	$6\frac{1}{4}$	$3\frac{1}{8}$
4	$6\frac{5}{16}$	$7\frac{3}{4}$	$9\frac{1}{2}$	$2\frac{1}{2}$	$7\frac{3}{8}$	$3\frac{11}{16}$
5 × 4	$6\frac{5}{16}$	$9\frac{5}{8}$	$9\frac{3}{4}$	3	$10\frac{3}{4}$	$5\frac{3}{8}$
5	$6\frac{1}{8}$	$9\frac{5}{8}$	10	4	$14\frac{3}{4}$	$7\frac{3}{8}$
6 × 4	$6\frac{3}{4}$	$9\frac{5}{8}$	$10\frac{7}{8}$	4 × 3	$4\frac{1}{2}$	$4\frac{5}{16}$
6 × 5	$6\frac{1}{8}$	$9\frac{5}{8}$	$10\frac{1}{2}$	5	15	$7\frac{1}{2}$
6	$8\frac{1}{4}$	$9\frac{1}{2}$	$12\frac{1}{4}$	5 × 4	$11\frac{3}{8}$	$5\frac{5}{16}$
8 × 6	$8\frac{7}{16}$	$10\frac{1}{2}$	$13\frac{3}{4}$	6	$16\frac{1}{4}$	$8\frac{1}{8}$
8	$11\frac{7}{8}$	14	17	6 × 4	$12\frac{3}{8}$	$5\frac{7}{16}$
				6 × 5	$13\frac{3}{8}$	$6\frac{1}{4}$

Dimensions for Special Elbows

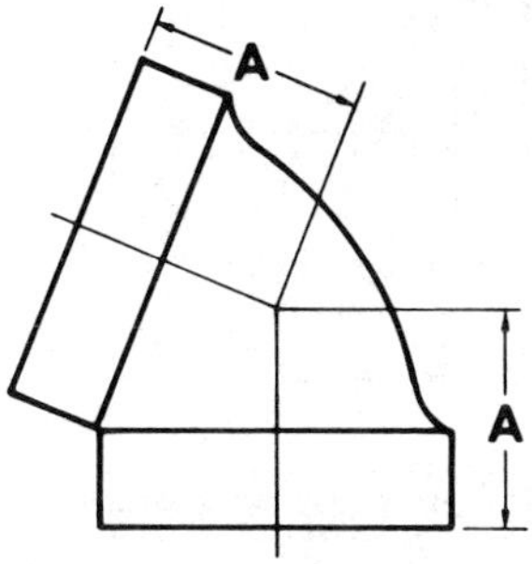

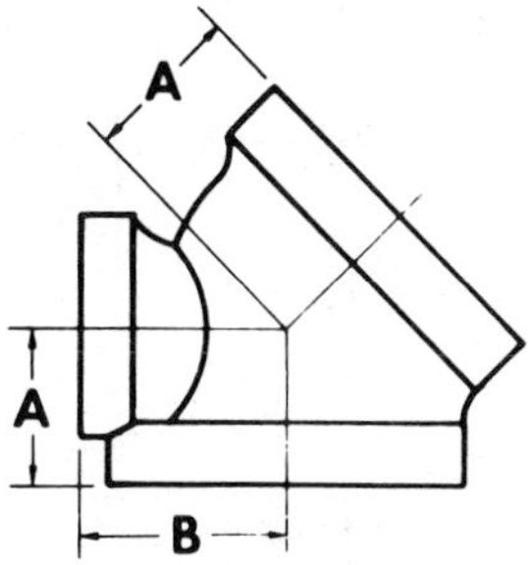

67 1/2° ELBOW		45° ELBOW WITH ANGLE HEEL OUTLET		
Size	A	Size	A	B
1 1/2 4	1 7/8 3 1/8	4	2 7/16	3 1/4

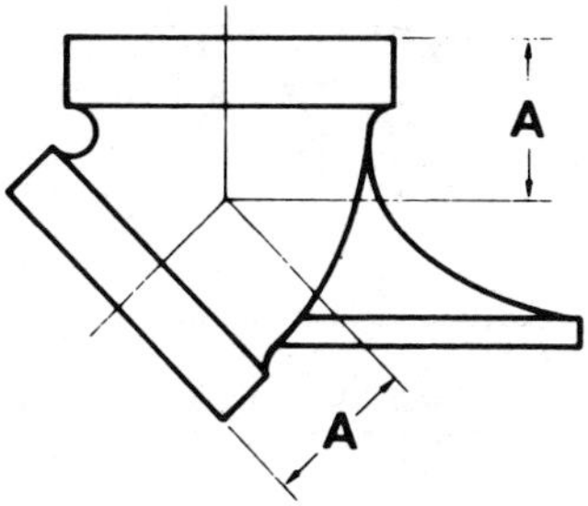

45° ELBOW WITH SHOE	
Size	A
2	1 5/8
3	2 1/4
4	2 7/16
5	2 11/16
6	3
8	3 1/2
10	4 7/16

Dimensions for Closet Elbows

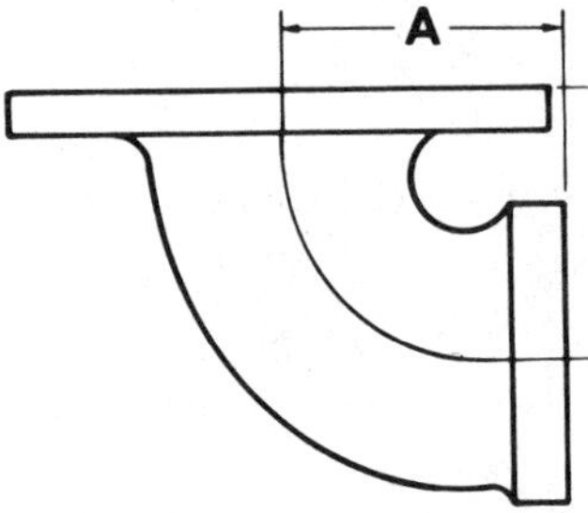

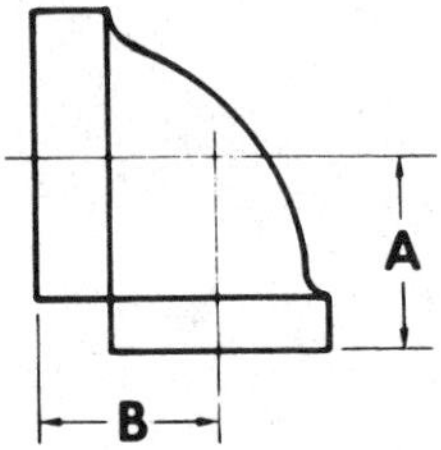

CLOSET ELBOW WITH FLANGE		
Size	A	B
3	5 7/8	4 5/8
4	6 3/4	5 1/2

REDUCING CLOSET ELBOW		
Size	A	B
5 × 4	4 7/16	3 13/16

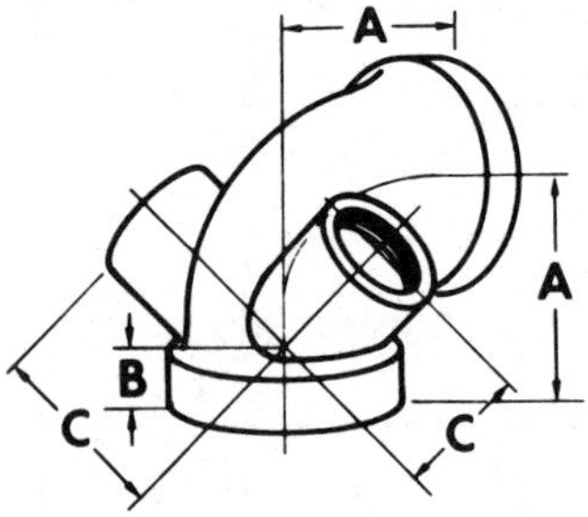

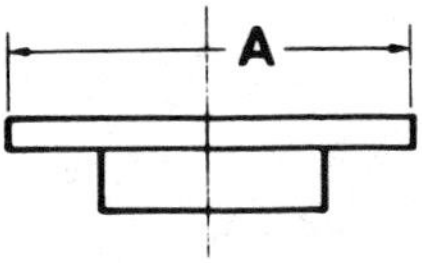

LONG-TURN 90° ELBOW WITH ANGLE INLETS ON BOTH SIDES			
Size	A	B	C
3	5 7/8	1 1/2	4 7/8
4	6 3/4	1 5/8	5 3/8
5	6 1/8	1 3/4	6 1/2
6	8 7/16	1 7/8	7 1/2

CLOSET FLANGE	
Size	A
4 × 7	7
4 × 10	10

Dimensions for TY's

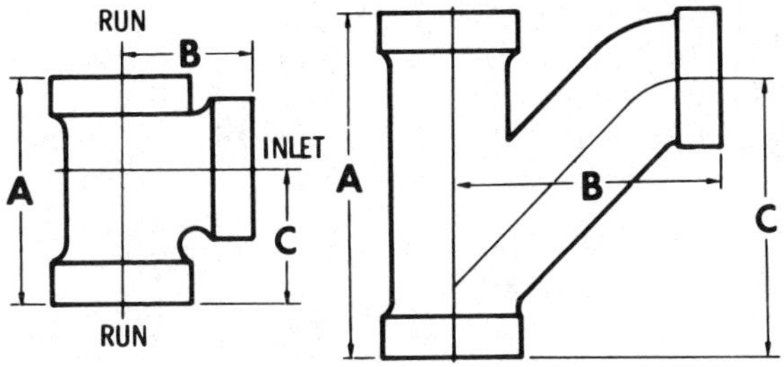

SHORT TY				LONG TY			
Size	A	B	C	Size	A	B	C
1¼	3 13/16	2 3/16	2 3/16	1¼	4¾	3⅝	3 7/16
1½	4 5/16	2½	2 7/16	1½	5 7/16	4 3/16	3⅞
2	5¾	3 3/16	3 3/16	2	6½	5⅛	4⅝
2½	6⅝	3⅞	3¾	2½	8¼	6 3/16	6 1/16
3	7¼	4 3/16	4 3/16	3	9	7⅛	6½
4	9⅛	5⅜	5⅝	4	10¾	8⅞	7⅝
5	10¼	6 3/16	6 3/16	5	13	10 3/16	9 5/16
6	12	7⅛	7⅛	6	14¼	10¾	10¼
7	13¾	8⅝	8½	7	16	12⅛	11¼
8	15¼	9⅛	9 5/16	8	17 7/16	13¼	11⅝
10	19½	12¼	12¼	10	22¾	16 7/16	15⅜
				12	26⅝	19	20

Dimensions for Reducing 45° TY's

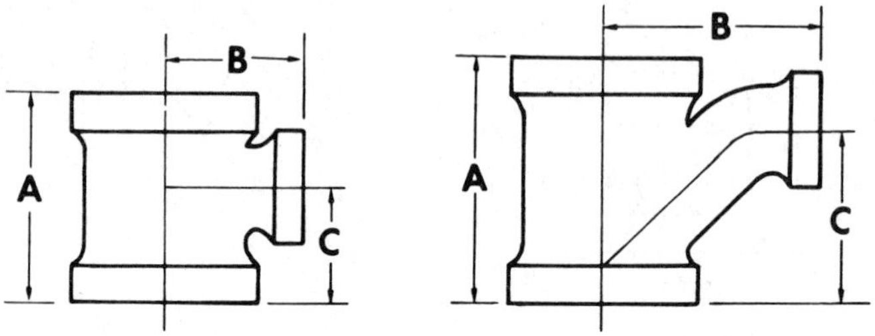

REDUCING SHORT TY				REDUCING LONG TY			
Size	A	B	C	Size	A	B	C
1 1/2 × 1 1/4	4 1/4	2 11/16	2 11/16	1 1/2 × 1 1/4	5 1/8	3 7/8	4 1/8
2 × 1 1/4	4 7/16	2 13/16	2 11/16	2 × 1 1/2	5 13/16	4 1/2	4 1/2
2 × 1 1/2	4 13/16	3	2 7/8	2 1/2 × 1 1/2	5 3/4	4 1/2	4 1/4
2 × 2 1/2	6 1/2	3 1/2	3 5/8	2 1/2 × 2	5 3/8	4 1/2	4 3/8
2 1/2 × 1 1/4	4 3/8	2 15/16	2 1/2	3 × 1 1/2	5 7/8	5 1/4	4 9/16
2 1/2 × 1 1/2	5 1/8	3 1/4	3	3 × 2	6 5/8	5 1/2	4 1/2
2 1/2 × 2	5 1/2	3 11/16	3 1/8	4 × 1 1/2	6 1/8	5 7/16	4 3/4
3 × 1 1/2	5 1/16	3 5/16	2 15/16	4 × 2	7	6 3/8	5 1/8
3 × 2	5 3/4	3 5/8	3 1/4	4 × 2 1/2	8 5/8	7 1/16	6 5/8
4 × 1 1/2	5 1/4	3 13/16	3	4 × 3	9 1/4	7 3/4	6 1/2
4 × 2	6	4 1/4	3 7/16	5 × 1 1/2	6 5/16	6	4 15/16
4 × 2 1/2	6 5/8	4 7/16	3 13/16	5 × 2	7 3/8	6 11/16	5 5/8
4 × 3	7 3/8	4 15/16	4 1/8	5 × 3	9 1/2	8 7/16	7
5 × 1 1/2	5 9/16	4 3/8	3 3/16	5 × 4	11 1/8	9 7/16	7 15/16
5 × 2	6 1/8	4 1/2	3 11/16	6 × 2	6 1/2	6 3/4	4 7/16
5 × 3	7 3/4	5 3/16	4 1/2	6 × 3	8 15/16	7 3/16	6 1/4
5 × 4	9 1/4	5 13/16	5 1/4	6 × 4	11 1/4	9 13/16	8 1/4
6 × 2	6 5/16	5 1/4	3 11/16	6 × 5	13	10 11/16	9 7/16
6 × 3	7 7/8	5 3/4	4 9/16	7 × 3	10 1/4	9 1/2	7 1/4
6 × 4	9 1/4	5 15/16	5 7/16	7 × 4	10 1/4	9 5/8	7
6 × 5	10 3/8	5 15/16	5 15/16	8 × 3	8 3/4	9 1/4	5 7/8
7 × 4	10 3/16	7 3/16	6 1/4	8 × 4	10 5/8	10	7 1/4
8 × 3	9 7/8	7 1/4	6 3/8	8 × 5	13 3/4	10 5/8	8 1/4
8 × 4	11 5/8	7 5/8	7 1/2	8 × 6	14 1/4	10 3/4	8 3/4
8 × 5	12 3/8	7 7/8	7 3/4	10 × 4	14	13	11
8 × 6	12 3/4	8 1/16	8 3/8	10 × 6	16 1/8	14 1/8	11 5/8
8 × 7	15	9	9	10 × 8	17 3/8	11 1/4	11 1/8
10 × 4	10 13/16	8 3/4	6 5/8	12 × 4	17	15 1/2	13
10 × 6	13 3/8	9 15/16	8 3/8	12 × 5	17	15 1/2	13
10 × 8	15 7/8	10	9 3/4				
12 × 4	15	9 1/2	10 1/2				
12 × 5	15	9 1/2	10				

Dimensions for 45° Y's

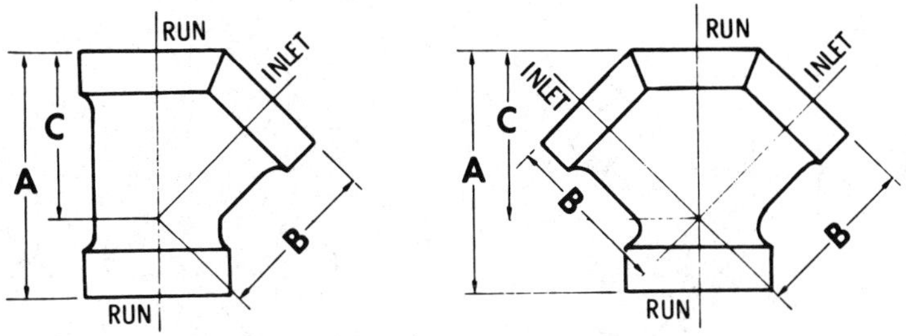

45° Y				DOUBLE 45° Y			
Size	A	B	C	Size	A	B	C
1	4	2 11/16	2 11/16	1 1/4	5 3/16	3 7/16	3 7/16
1 1/4	5 3/16	3 7/16	3 7/16	1 1/2	5 1/4	3 7/16	3 7/16
1 1/2	5 1/4	3 7/16	3 7/16	2	6 1/4	4 1/8	4 1/8
2	6 1/4	4 1/8	4 1/8	2 1/2	8	5 1/4	5 1/4
2 1/2	8	5 1/4	5 1/4	3	8 3/8	6	6
3	8 3/8	6	6	4	9 5/8	7	7
4	9 5/8	7	7	5	11 3/8	8 3/8	8 3/8
5	11 3/8	8 3/8	8 3/8	6	13 1/4	10	10
6	13 1/4	10	10	7	15 1/4	11 1/4	11 1/4
7	15 1/4	11 1/4	11 1/4	8	16 1/2	12 3/8	12 3/8
8	16 1/2	12 3/8	12 3/8	10	20 1/4	15 1/4	15 1/4
10	20 1/4	15 1/4	15 1/4	12	24 1/4	19 5/8	19 5/8
12	24 1/4	19 5/8	19 5/8				

Dimensions for Reducing 45° Y's

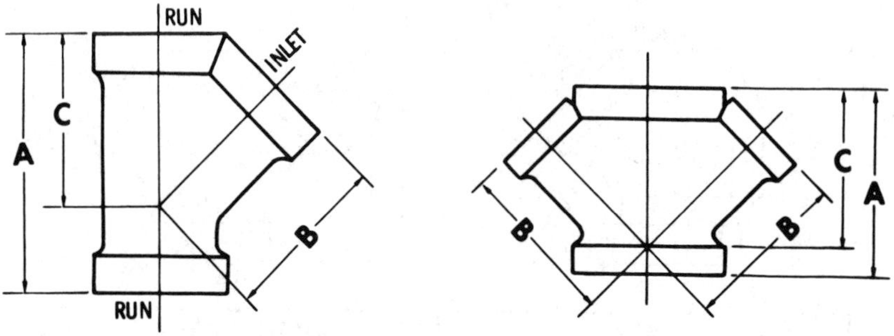

REDUCING 45° Y			
Size	A	B	C
1 1/2 × 1 1/4	5 1/4	3 7/16	3 7/16
2 × 1 1/2	5 15/16	4 1/4	4 1/4
2 1/2 × 1 1/2	6 1/4	4 5/8	4 9/16
2 1/2 × 2	6 3/8	4 3/4	4 5/8
3 × 1 1/2	6 5/8	5 1/16	4 11/16
3 × 2	6 3/4	5 1/4	5 1/8
3 × 2 1/2	8	5 13/16	5 11/16
4 × 1 1/2	7 5/8	5 5/8	5 7/16
4 × 2	6 5/8	5 7/8	5 5/8
4 × 2 1/2	7 1/4	5 7/8	5 3/4
4 × 3	8 1/2	6 5/8	6 3/8
5 × 2	7 1/8	6 3/4	6
5 × 3	8 3/4	7 1/2	7 1/8
5 × 4	10 5/16	8	7 7/8
6 × 2	7	7 1/2	6 5/8
6 × 3	8 13/16	8 1/4	7 1/2
6 × 4	10 1/2	9	8 7/8
6 × 5	11 1/2	9 1/8	9
7 × 3	10	9 5/8	8 7/8
7 × 4	10 1/4	10 1/2	9 5/8
7 × 5	13 1/2	10 1/2	10 1/2
7 × 6	15 3/8	11 9/16	11 3/8
8 × 3	9	10	8 7/8
8 × 4	10 3/8	10	8 7/8
8 × 5	11 3/8	10 5/8	9 3/4
8 × 6	14	12	11 1/2
8 × 7	16 6/8	12 7/16	12 7/16
10 × 4	11 7/8	12 5/8	11 5/8
10 × 5	13 3/4	12 3/4	12
10 × 6	13 7/8	13 5/8	13
10 × 7	17 1/4	14	14
10 × 8	17 1/2	14 3/8	14 3/8
12 × 4	15	13 3/8	11 3/8
12 × 5	15	14	12 1/4
12 × 6	15	14 5/8	13 1/2

REDUCING DOUBLE 45° Y			
Size	A	B	C
1 1/2 × 1 1/4	5 1/4	3 7/16	3 7/16
2 × 1 1/2	5 15/16	4 1/4	4 1/4
2 1/2 × 1 1/2	6 1/4	4 5/8	4 9/16
2 1/2 × 2	6 3/8	4 3/4	4 5/8
3 × 1 1/2	6 5/8	5 1/16	4 11/16
3 × 2	6 3/4	5 1/4	5 1/8
4 × 2	6 5/8	5 7/8	5 5/8
4 × 3	8 1/2	6 5/8	6 3/8
5 × 2	7 1/8	6 3/4	6
5 × 3	8 3/4	7 1/2	7 1/8
5 × 4	10 5/16	8	7 7/8
6 × 2	7	7 1/2	6 5/8
6 × 3	8 13/16	8 1/4	7 1/2
6 × 4	10 1/2	9	8 7/8
6 × 5	11 1/2	9 1/8	9
7 × 4	10 1/4	10 1/2	9 5/8
7 × 5	13 1/2	10 1/2	10 1/2
7 × 6	15 3/8	11 9/16	11 3/8
8 × 3	9	10	8 7/8
8 × 4	10 3/8	10	8 7/8
8 × 5	11 3/8	10 5/8	9 3/4
8 × 6	14 1/8	12 5/16	11 13/16
8 × 7	16 3/8	12 7/16	12 7/16
10 × 4	11 7/8	12 5/8	11 5/8
10 × 5	13 3/4	12 3/4	12
10 × 6	13 7/8	13 5/8	13
10 × 7	17 1/4	14	14
10 × 8	17 1/2	14 3/8	14 3/8

Dimensions for 60° Y's and T's

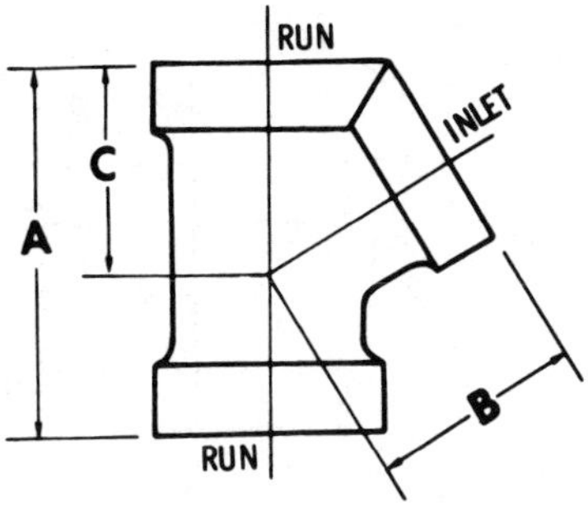

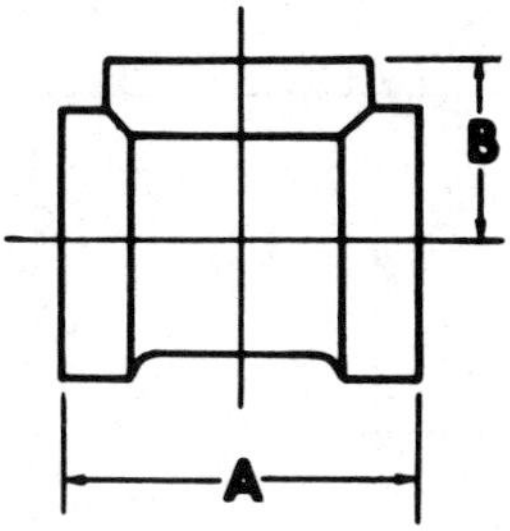

60° Y				TEE		
Size	A	B	C	Size	A	B
1¼	4⅞	2¾	2¾	1¼	3¾	1⅞
1½	5¼	3 1/16	2⅞	1½	4⅛	2
2	6¼	3⅜	3¾	2	5⅛	2½
2½	6½	4⅛	4⅛	2½	5⅛	2 9/16
3	7¼	4⅞	4⅞	3	6⅜	3¼
4	8 7/16	6	5¾	4	8	4⅛
5	10¼	7	7	5	9⅛	4 9/16
6	11⅝	7⅞	7⅞	6	10⅜	5 3/16
8	16½	10¾	10¾	7	11⅝	5 13/16
				8	13	6 7/16
				10	15	7½
				12	18	9

Dimensions for 60° Y's and T's

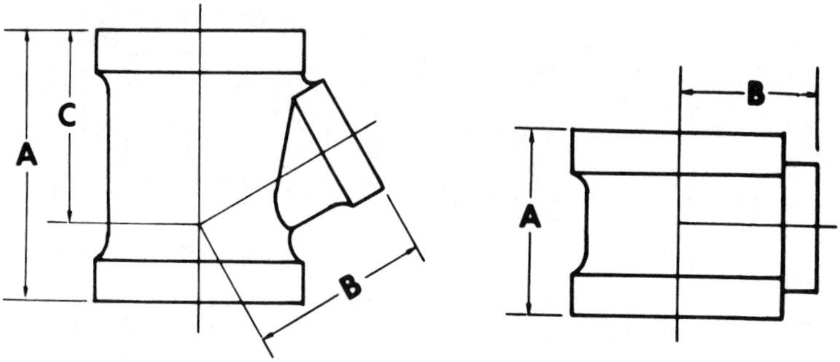

REDUCING 60° Y				REDUCING TEE		
Size	A	B	C	Size	A	B
1 1/2 × 1 1/4	4 3/4	2 7/8	2 7/8	1 1/2 × 1 1/4	3 7/8	1 15/16
2 × 1 1/2	6 1/4	2 7/16	3 3/4	2 × 1 1/2	4 3/8	2 1/2
2 1/2 × 1 1/2	6 5/8	4 1/4	4 1/4	2 1/2 × 1 1/2	4 1/2	2 3/4
2 1/2 × 2	6 7/8	4 3/8	4 3/8	2 1/2 × 2	5 3/16	2 3/4
3 × 1 1/2	6 3/8	4 1/2	4 3/16	3 × 1 1/2	5 1/2	3 1/8
3 × 2	6 7/16	4 1/8	3 15/16	3 × 2	5 1/2	3 1/8
4 × 1 1/2	7 1/16	4 5/8	5 1/16	4 × 2	4 7/8	3 5/8
4 × 2	7 1/16	4 5/8	5 1/16	4 × 3	6 3/4	3 7/8
4 × 3	7 3/4	5 1/2	5 1/4	5 × 2	6 7/8	4 7/16
5 × 2	7	5 3/4	4 7/8	5 × 3	6 7/8	4 7/16
5 × 3	8 1/2	6	5 9/16	5 × 4	8 1/8	4 9/16
5 × 4	9 1/4	7	7 7/8	6 × 2	8 1/8	5 1/8
6 × 2	7 1/2	6 1/4	6 1/8	6 × 3	8 1/8	5 1/8
6 × 3	9 3/4	7 1/4	7	6 × 4	8 1/8	5 1/8
6 × 4	9 3/4	7 1/4	7	6 × 5	9 1/2	5 3/16
6 × 5	11	7 1/2	7 3/4	7 × 4	8 11/16	6
7 × 4	10 3/4	8 5/8	7 5/16	8 × 4	8 13/16	6
8 × 3	8 3/4	8	6 1/2	8 × 5	10	6
8 × 4	9 3/4	8 11/16	7 3/8	8 × 6	10 1/4	6
8 × 5	12 3/8	9 3/8	8 7/16	8 × 7	12 1/4	6 3/4
8 × 6	12 3/8	9 3/8	8 7/16	10 × 5	10 5/16	7 1/2
8 × 7	14 5/8	10	10	10 × 6	11 3/8	7 7/16
				10 × 7	12 5/8	7 7/8
				10 × 8	13 5/8	8
				12 × 8	15	9
				12 × 10	16 1/2	9

Dimensions for Crosses

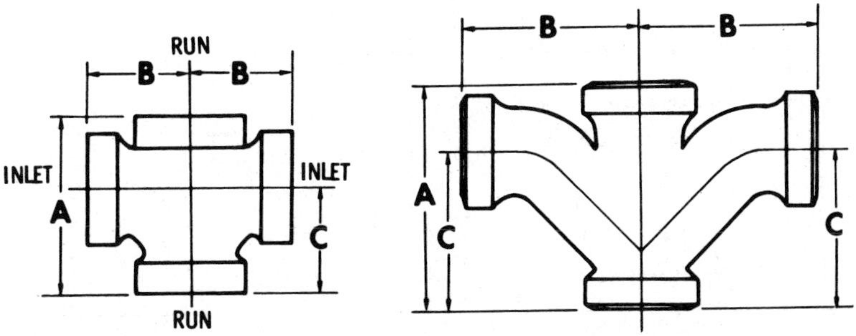

SHORT CROSS				LONG CROSS			
Size	A	B	C	Size	A	B	C
1 1/4	3 13/16	2 3/16	2 3/16	1 1/4	4 3/4	3 5/8	3 7/16
1 1/2	4 5/16	2 1/2	2 7/16	1 1/2	5 7/16	4 3/16	3 7/8
2	5 3/4	3 1/16	3 3/16	2	6 1/2	5 1/8	4 5/8
2 1/2	6 5/8	3 7/8	3 3/4	2 1/2	8 1/4	6 3/16	6 1/16
3	7 1/4	4 3/16	4 3/16	3	9	7	6 1/2
4	9 1/8	5 3/8	5 3/8	4	10 3/4	8 5/8	7 5/8
5	10 1/4	6 3/16	6 3/16	5	13	10 3/16	9 5/16
6	12	7 1/8	7 1/8	6	14 1/4	10 3/4	10 1/4
7	13 3/4	8 5/8	8 1/2	7	16	12 1/8	11 1/4
8	15 1/4	9 1/8	9 5/16	8	17 7/16	13 1/4	11 5/8
10	19 1/2	12 1/4	12 1/4	10	22 3/4	16 7/16	15 3/8
				12	26 5/8	19	20

Dimensions for Reducing Crosses

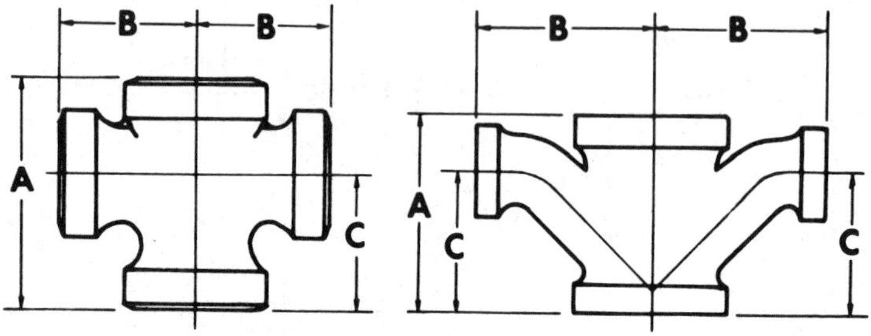

REDUCING SHORT CROSS				REDUCING LONG CROSS			
Size	A	B	C	Size	A	B	C
1½ × 1¼	4⅜	2¾	2¾	1½ × 1¼	5⅛	3⅞	4⅛
2 × 1¼	4 7/16	2 13/16	2 11/16	2 × 1½	6⅜	4 5/16	5
2 × 1½	4 15/16	3	2⅞	2½ × 1½	5¾	4½	4¼
2½ × 1½	5½	3 11/16	3⅛	2½ × 2	5⅜	4½	4⅜
2½ × 2	5½	3 11/16	3⅛	3 × 1½	5⅞	5¼	4 9/16
3 × 1½	5 1/16	3 5/16	2 15/16	3 × 2	6⅝	5½	4½
3 × 2	5¾	3⅝	3¼	4 × 2	7	6⅜	5⅛
4 × 2	6	4¼	3 7/16	4 × 3	9¼	7¾	6½
4 × 3	7⅜	4 15/16	4⅛	5 × 2	7⅜	6 11/16	5⅝
5 × 2	6⅛	4½	3 11/16	5 × 3	9½	8 7/16	7
5 × 3	7¾	5 3/16	4½	5 × 4	11⅛	9 7/16	7 15/16
5 × 4	9¼	5 13/16	5¼	6 × 2	6½	6¾	4 7/16
6 × 2	6¼	5⅛	3⅝	6 × 3	8⅞	7¼	6¼
6 × 3	7⅞	5¾	4 9/16	6 × 4	11¼	9 13/16	8¼
6 × 4	9¼	5 15/16	5 7/16	6 × 5	13	10 11/16	9 7/16
6 × 5	10⅜	5 15/16	5 15/16	7 × 4	10¼	9½	7¼
7 × 4	10 3/16	7 5/16	6¼	8 × 3	8¾	9¼	5⅞
8 × 3	9⅞	7¼	6⅜	8 × 4	10⅝	10	7¼
8 × 4	11⅝	7⅝	7½	8 × 5	13¾	10⅝	8¼
8 × 5	12⅜	7⅞	7¾	8 × 6	14¼	10¾	8¾
10 × 4	10 13/16	8¾	6⅝	10 × 4	14	13	11
10 × 6	13⅜	9 15/16	8⅜	10 × 6	16⅛	14⅛	11⅝
10 × 8	15⅞	10	9¾	10 × 8	18	14½	13

Dimensions for Miscellaneous Fittings

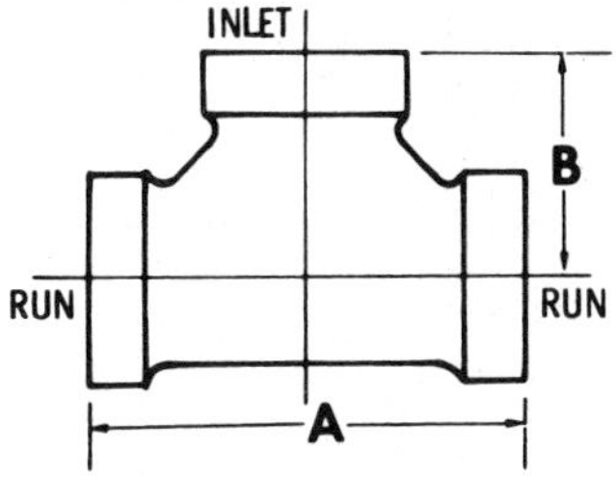

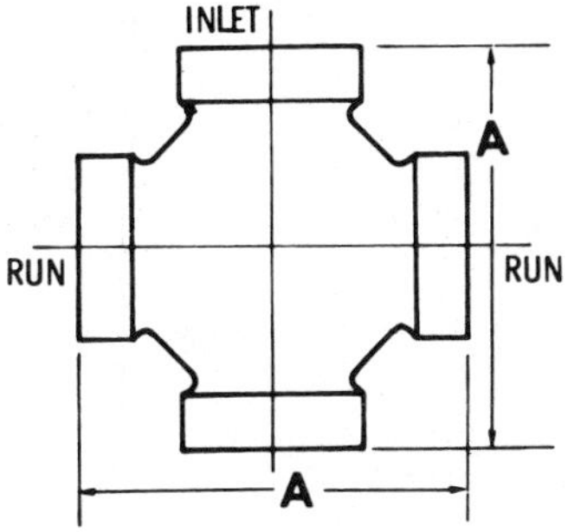

BASIN TEE			BASIN CROSS		
Size	A	B	Size	A	B
1 1/4	4 3/4	2 3/8	1 1/2	5 5/8	5 5/8
1 1/2	5 5/8	2 3/4	2	7	7
1 1/2 × 1 1/4	5 1/8	2 9/16	2 × 1 1/2	6 3/4	6 3/4
2	7	3 1/2			
2 × 1 1/2	6 3/4	3 3/8			
2 1/2	8 1/4	4 1/4			
2 1/2 × 1 1/4	5 3/4	3 1/2			

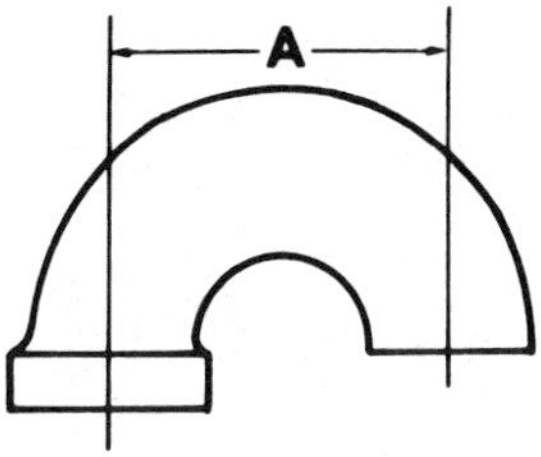

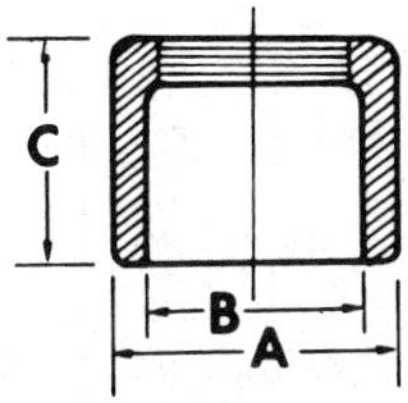

AIR CAPPING		ROOF CONNECTION			
Size	A	Size	A	B	C
2	5	1 1/2	2 7/8	2 1/4	3
3	5 1/2	2	3 1/2	3	3 1/2
4	7	3	5	4 1/4	3 3/4
5	8	4	5 7/8	5 1/8	5
		5	7 3/8	6 1/2	6
		6	8 1/4	7 5/8	6 7/8

Dimensions for Long-Turn TY

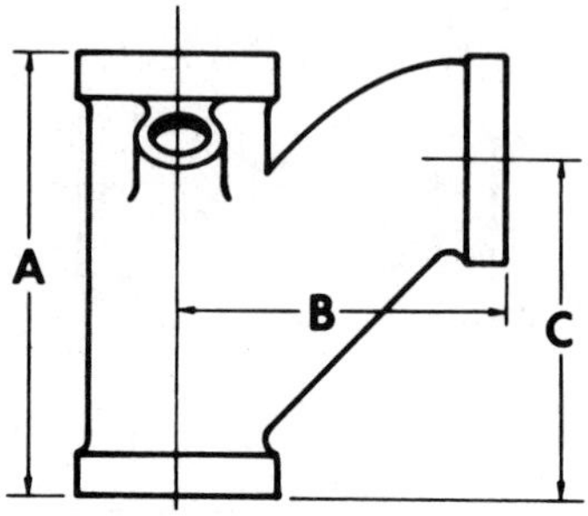

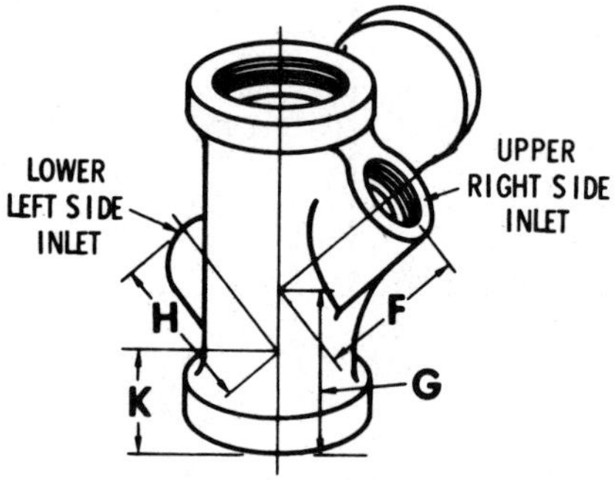

LONG-TURN TY (Closet Tee) with either Right or Left Side Inlet							
Size	A	B	C	F	G	H	K
3	9	7 1/8	6 1/2	4 15/16	3 1/8	4 15/16	2 5/16
4	10 3/4	8 7/8	7 5/8	5 3/4	4 3/8	5 3/4	1 7/8
5 × 4	11 1/8	9 7/16	7 15/16	6 1/4	3 3/4	6 1/2	1 5/8
6 × 4	11 1/4	9 3/16	8 1/4	7 1/4	3 1/2	7 1/4	1 1/8

Note—The 90° inlet of closet T's is tapped pitched 1/4 in. to the foot. The size of the side and top inlets is 2 inches.

Dimensions for Long-Turn TY

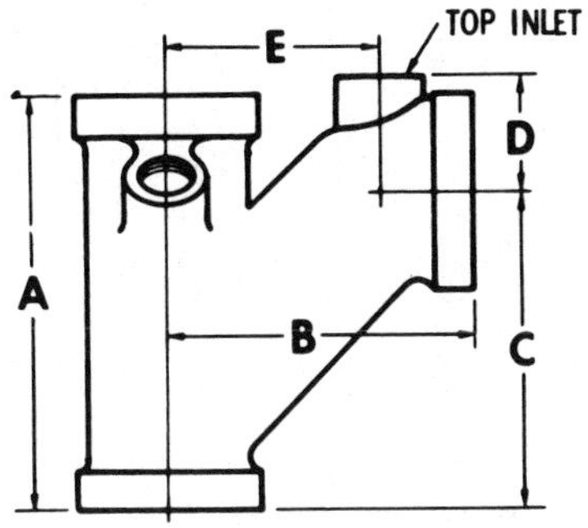

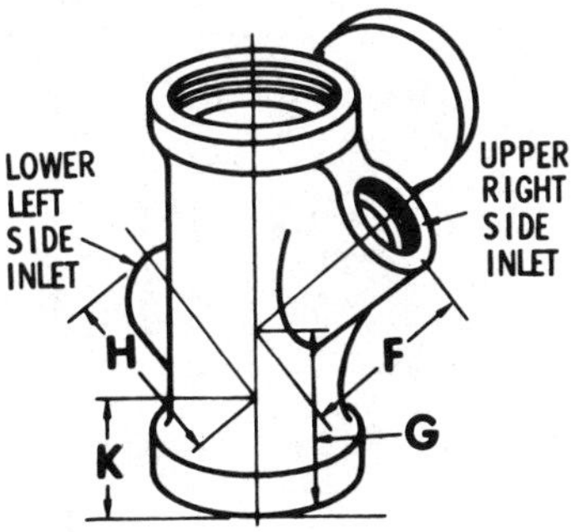

LONG-TURN TY (Closet Tee) with Either Right or Left Side Inlet and Top Inlet									
Size	A	B	C	D	E	F	G	H	K
3	9	7 1/8	6 1/2	2 3/4	4 1/8	4 15/16	3 1/8	4 15/16	2 5/16
4	10 3/4	8 7/8	7 5/8	3 3/8	5 3/16	5 3/4	4 3/8	5 3/4	1 7/8
5 × 4	11 1/8	9 7/16	7 15/16	3 3/8	6 3/16	6 1/4	3 3/4	6 1/2	1 5/8
6 × 4	11 1/4	9 3/16	8 1/4	3 3/8	7 1/8	7 1/4	3 1/2	7 1/4	1 1/8

Note—The 90° inlet of closet T's is tapped pitched 1/4-in. to the foot. The size of the side and top inlets is 2 inches.

Dimensions for Miscellaneous Fittings

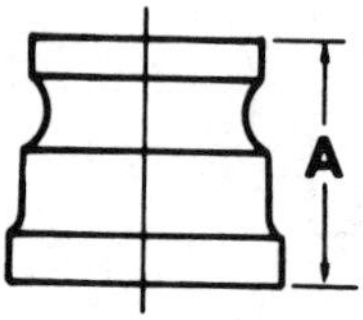

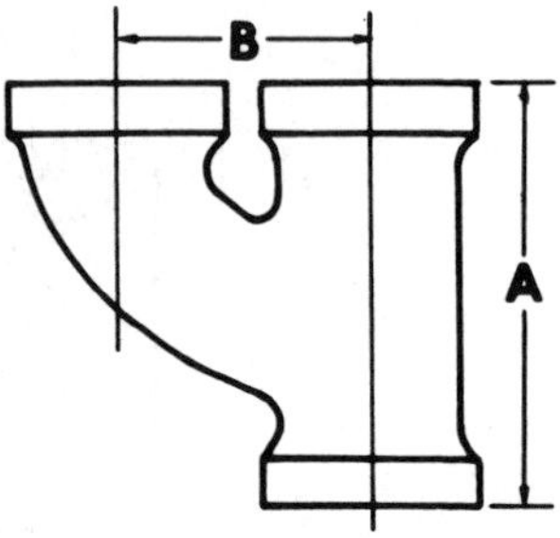

TUCKER CONNECTION		SPECIAL UPRIGHT Y		
Size	A	Size	A	B
2	$3\frac{5}{8}$	$1\frac{1}{4}$	$6\frac{1}{4}$	$3\frac{1}{4}$
3	$4\frac{5}{8}$	$1\frac{1}{2}$	$6\frac{13}{16}$	$3\frac{5}{8}$
4	$6\frac{1}{4}$	2	$6\frac{5}{8}$	4
5	$6\frac{1}{2}$	$2 \times 1\frac{1}{2}$	$6\frac{5}{8}$	4
6	$6\frac{11}{16}$	3	$8\frac{1}{4}$	$5\frac{1}{16}$
8	$6\frac{13}{16}$	3×2	$6\frac{1}{4}$	4
10	$7\frac{1}{8}$	4	$10\frac{1}{4}$	$6\frac{3}{8}$
12	$9\frac{1}{8}$	4×2	$6\frac{11}{16}$	$4\frac{3}{8}$

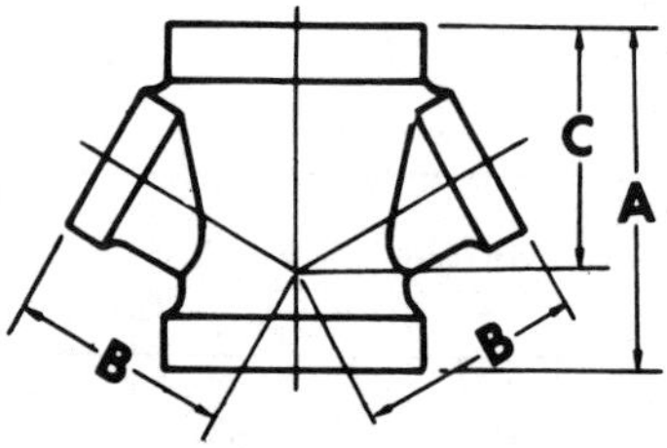

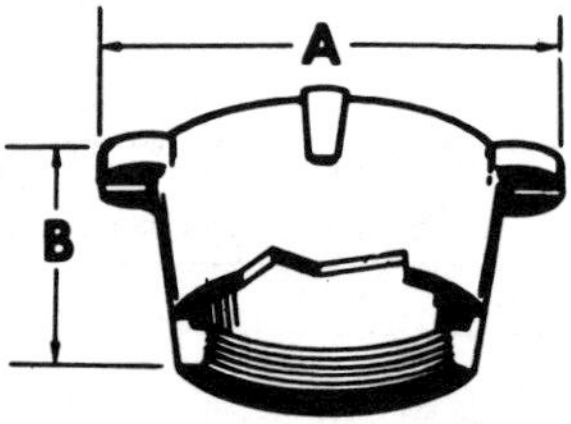

REDUCING DOUBLE 60° Y				SINK COUPLING		
Size	A	B	C	Size	A	B
$2 \times 1\frac{1}{2}$	$5\frac{5}{16}$	$3\frac{1}{4}$	$3\frac{1}{4}$	$1\frac{1}{4}$	4	$2\frac{1}{8}$
				$1\frac{1}{2}$	4	$2\frac{1}{8}$
				2	4	$2\frac{1}{8}$

Dimensions for Increasers and Offsets

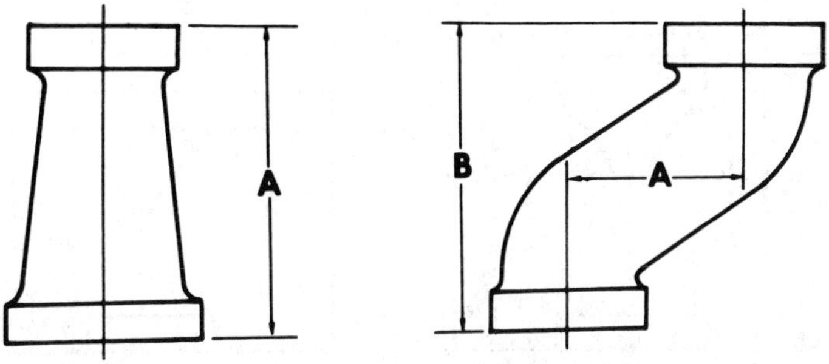

INCREASER		OFFSET			
Size	A	Size	To Offset	A	B
2 × 1½	9	2	4	4	7½
2½ × 2	9	2	6	6	10 7/16
3 × 2	9	2	8	8	12 7/16
3 × 2½	9	2	10	10	14 7/16
4 × 2	9	3	4	4	9¼
4 × 3	9	3	6	6	10½
5 × 2	9	3	8	8	13
5 × 3	9	3	10	10	14¾
5 × 4	9	3	12	12	17⅛
6 × 4	9	4	4	4	10⅛
6 × 5	9	4	6	6	11⅝
7 × 6	9	4	8	8	15⅜
8 × 4	9	4	10	10	16⅛
8 × 6	9	4	12	12	18⅛
10 × 6	9	5	6	6	13
10 × 8	9	5	8	8	15
12 × 10	9	5	10	10	17
		5	12	12	19
		6	6	6	13⅝
		6	8	8	15⅛
		6	10	10	17⅞
		6	12	12	19⅞

Dimensions for Special TY's

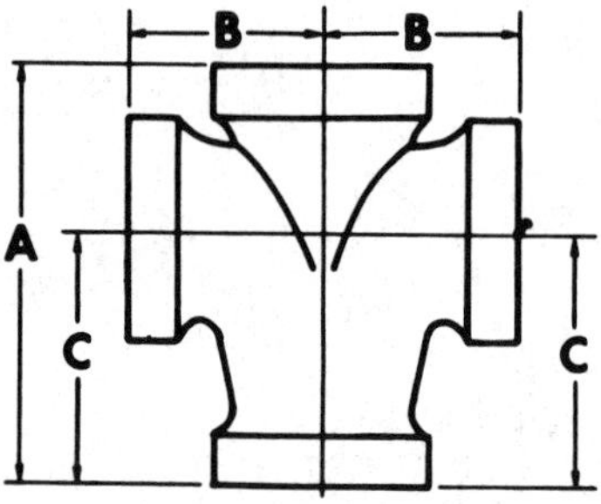

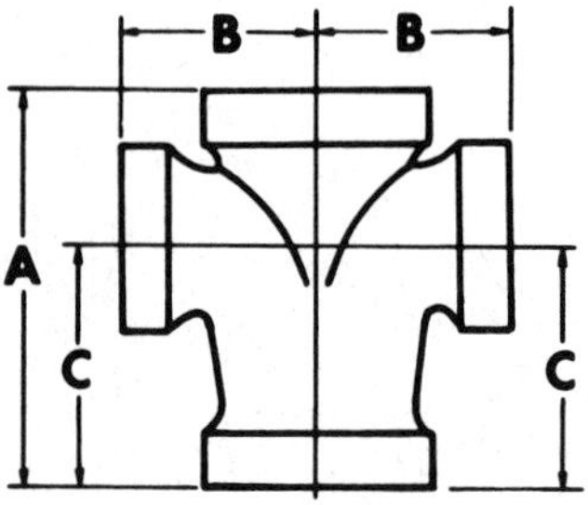

SPECIAL DOUBLE TY (Short Branch, Long Body)				SPECIAL REDUCING DOUBLE TY (Short Branch, Long Body)			
Size	A	B	C	Size	A	B	C
1 1/4	4 13/16	2 1/4	2 15/16	2 × 1 1/2	5 3/4	3	3 1/2
1 1/2	5 3/8	2 1/2	3 5/16	2 1/2 × 1 1/2	5 3/4	3 11/32	3 13/16
2	6 3/8	3 1/16	3 11/16	3 × 1 1/2	6	3 5/8	3 5/8
2 1/2	7 3/16	3 11/16	4 7/16	3 × 2	7 5/8	3 3/4	5
3	9	4 3/8	5 3/4				

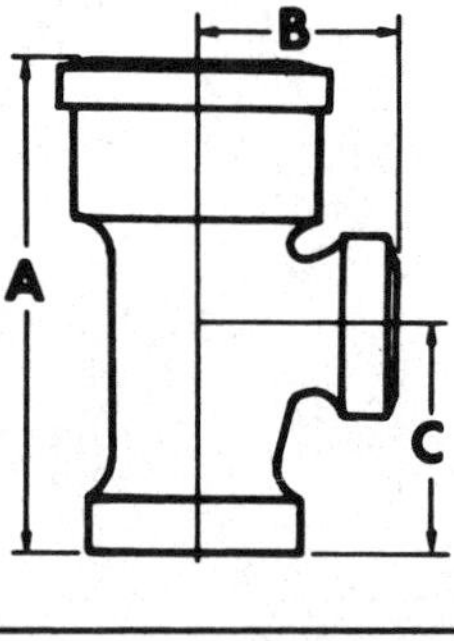

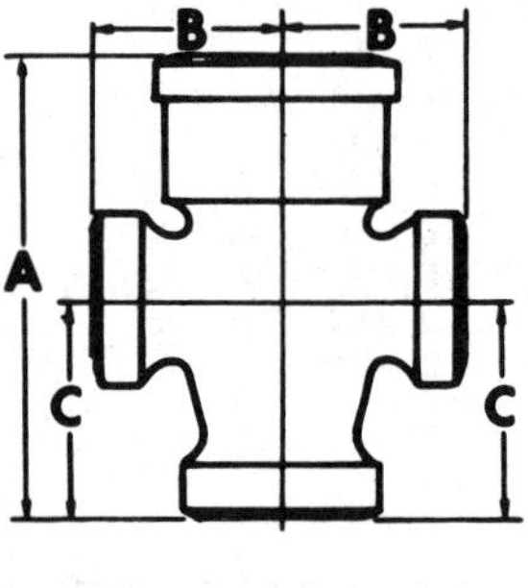

SPECIAL TY (With Hub for Soil Pipe Connection)							SPECIAL DOUBLE TY (With Hub for Soil Pipe Connection)						
Hub	End	Run	Br'ch	A	B	C	Hub	End	Run	Br'ch	A	B	C
1 1/2"	W.I.	1 1/4	1 1/4	5 3/8	2 5/16	2 5/16	2"	Soil	2	1 1/2	7 1/4	3	3 5/8
2"	W.I.	1 1/2	1 1/2	6 1/4	3 3/8	3 3/8							
2"	Soil	1 1/2	1 1/2	7	3 3/8	3 3/8							
2"	Soil	2	2	7 1/2	3 13/16	3 13/16							

Dimensions for Special TY's and Couplings

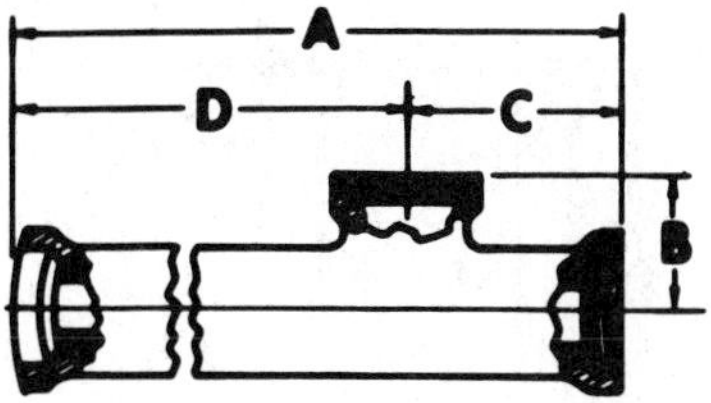

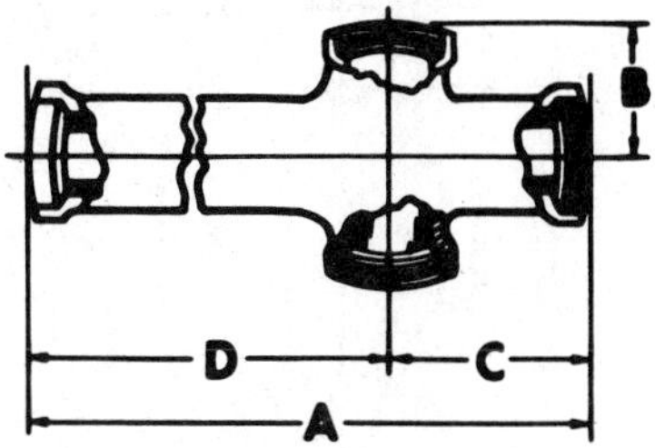

SPECIAL LONG TY					SPECIAL LONG DOUBLE TY				
Size	A	B	C	D	Size	A	B	C	D
$1\frac{1}{2}$	24	$2\frac{1}{2}$	4	20	$1\frac{1}{2}$	24	$2\frac{1}{2}$	4	20
$1\frac{1}{2}$	34	$2\frac{1}{2}$	4	30	$1\frac{1}{2}$	34	$2\frac{1}{2}$	4	30

COUPLING	
Size	A
$1\frac{1}{4}$	3
$1\frac{1}{2}$	$3\frac{7}{16}$
2	$3\frac{1}{2}$
$2\frac{1}{2}$	4
3	$4\frac{1}{4}$
4	$4\frac{1}{2}$
5	$4\frac{3}{4}$
6	$5\frac{1}{4}$
7	$5\frac{1}{2}$
8	$5\frac{5}{8}$
10	$6\frac{1}{4}$

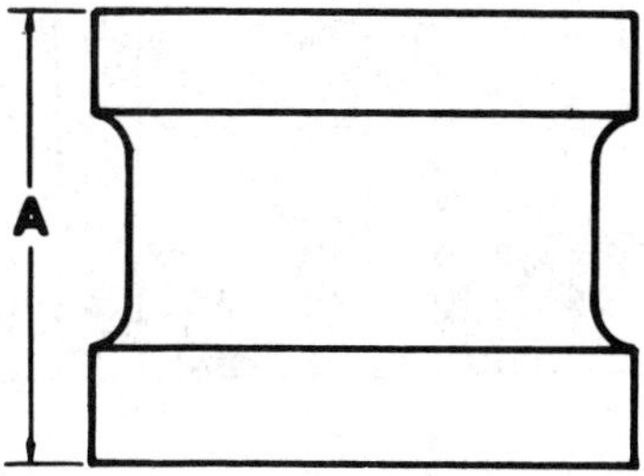

Dimensions for Traps

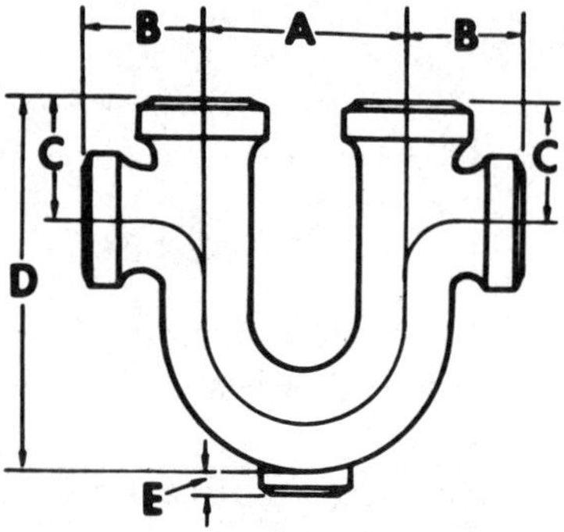

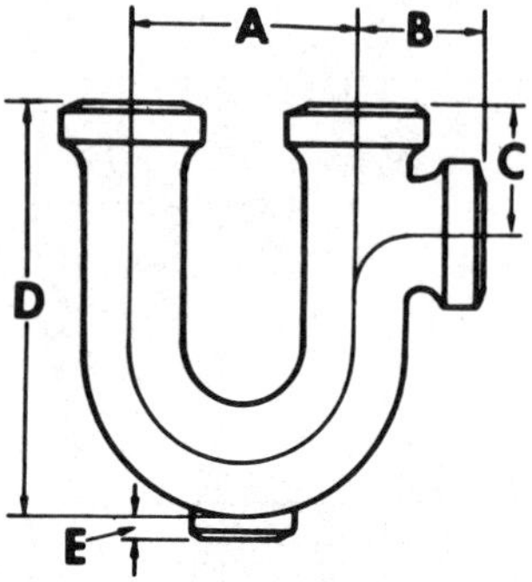

RUNNING TRAP						1/2 S TRAP					
Size	A	B	C	D	E	Size	A	B	C	D	E
1 1/4	3	2 7/8	1 7/8	5 5/8	1/4	1 1/4	3	2 7/8	1 7/8	5 5/8	1/4
1 1/2	4	3 7/16	1 15/16	7	1/4	1 1/2	4	3 7/16	1 15/16	7	1/4
2	4	3 1/16	2 9/16	7 1/8	7/16	2	4	3 1/16	2 9/16	7 1/8	7/16
3	6 7/16	3 3/4	3 1/2	11 1/2	1/4	3	6 1/2	3 1/8	3 1/2	11 1/2	1/4
4	7 1/2	5 1/4	4	13 7/8	1/4	4	7 1/2	5 1/4	4	13 3/4	1/4
5	8 7/16	6	4 3/4	15 7/8	1/4	5	8 7/8	6 1/4	4 13/16	16	1/4
6	10	6 3/4	5 11/16	19 1/4	1/4	6	10 1/8	7	5 3/4	19 1/16	7/16
7	11 1/4	7 3/4	6 1/8	21	1/4	7	11 1/4	7 3/4	6 1/8	21	1/4
8	12 3/8	9 1/8	6 5/8	23 3/4	1/4	8	12 3/8	9 1/8	6 5/8	23 3/4	1/4
10	14 1/4	10 3/16	7 1/2	27 1/4	1/4	10	14 1/4	10 3/16	7 1/2	27 1/4	1/4
12	16 1/2	11 1/2	8 1/2	31 3/4	3/8						

Dimensions for Traps

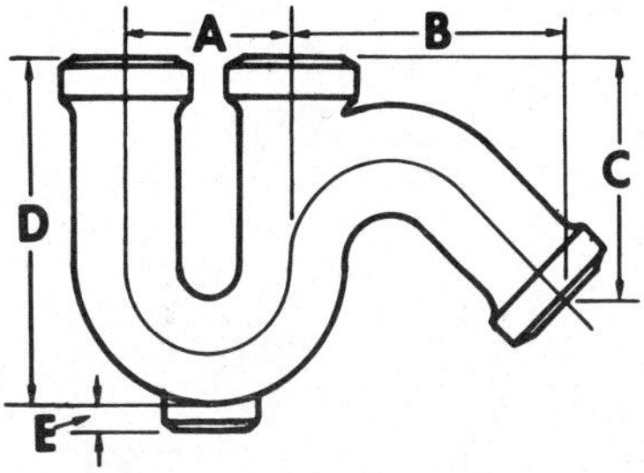

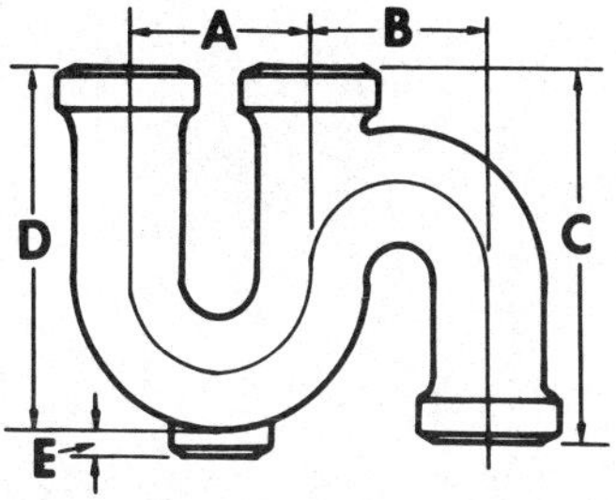

¾ S TRAP						S TRAP					
Size	A	B	C	D	E	Size	A	B	C	D	E
1½	3	2¼	4	5⅜	⅜	1½	3	1⅞	5⅝	5⅜	⅜
2	4 1/16	5⅝	5⅝	8⅛	⅜	2	4¼	4 1/16	8⅛	8	⅜
3	6½	7¼	5⅝	11½	⅜	3	8 3/16	2 13/16	11	10⅝	⅜
4	7 11/16	8¼	6⅞	13⅝	¼	4	10⅛	3 5/16	13¾	13	⅜
5	8⅜	11⅝	10¼	15¾	¼	5	12 21/32	4 7/32	16⅞	15⅞	⅜
6	10⅛	13⅞	11⅜	19⅛	¼	6	15	5	19¾	19¼	⅜

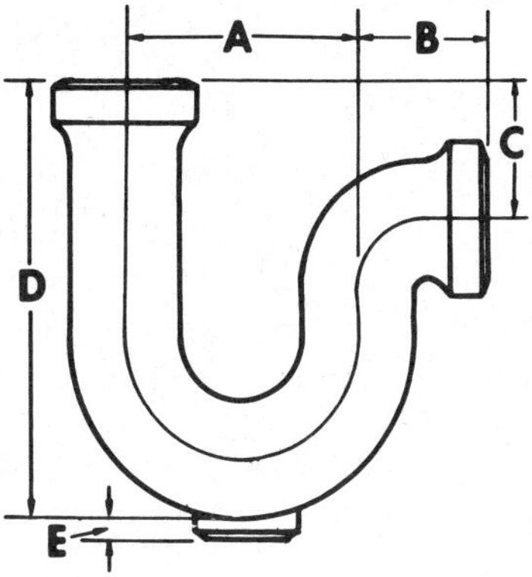

½ S TRAP, NO VENT				
Size	A+B	C	D	E
1½	7 7/16	1 15/16	7	¼
2	8⅜	2⅝	9⅜	¼
3	10⅜	3½	11½	¼

Dimensions for Drainage Fittings

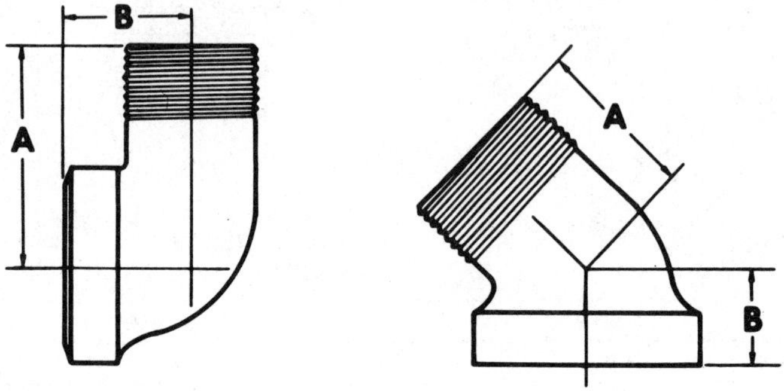

90° STREET ELBOW			45° STREET ELBOW		
Size	A	B	Size	A	B
1 1/4	2 5/8	1 3/4	1 1/4	1 3/4	1 5/16
1 1/2	3	1 13/16	1 1/2	2	1 1/4
2	3	2 1/4	2	2 1/4	1 5/8

Tappings and Dimensions

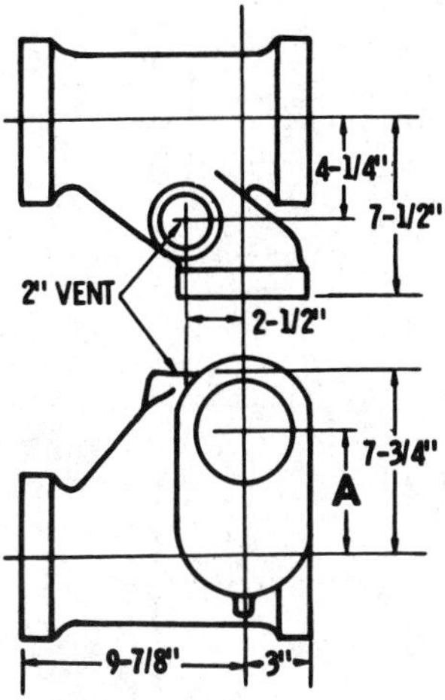

RIGHT-HAND

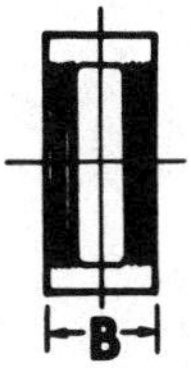

OFFSET COUPLING

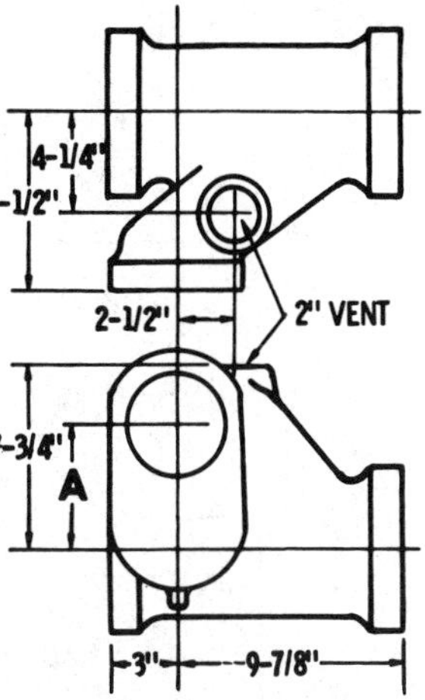

LEFT-HAND

4″ × 4″ and 5″ × 4″	
Tapping No.	Dimension A
1	$5\frac{1}{4}$
2	$4\frac{7}{8}$
3	$4\frac{1}{2}$
4	$4\frac{1}{8}$
5	$3\frac{3}{4}$
6	$3\frac{3}{8}$
7	3
8	$2\frac{5}{8}$
9	$2\frac{1}{4}$
10	$1\frac{7}{8}$
11	$1\frac{1}{2}$
12	$1\frac{1}{8}$

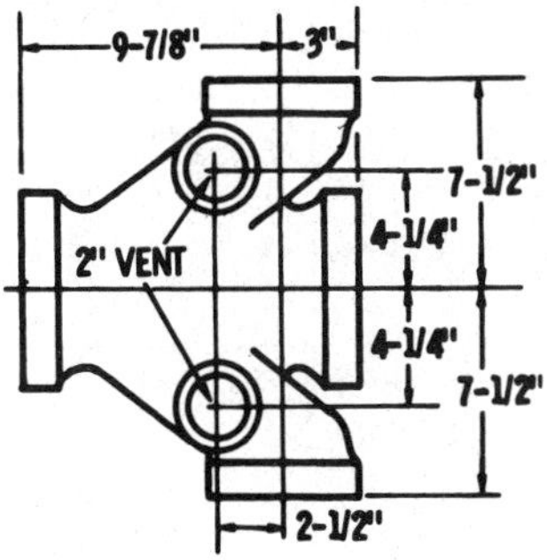

DOUBLE

Tappings and Dimensions

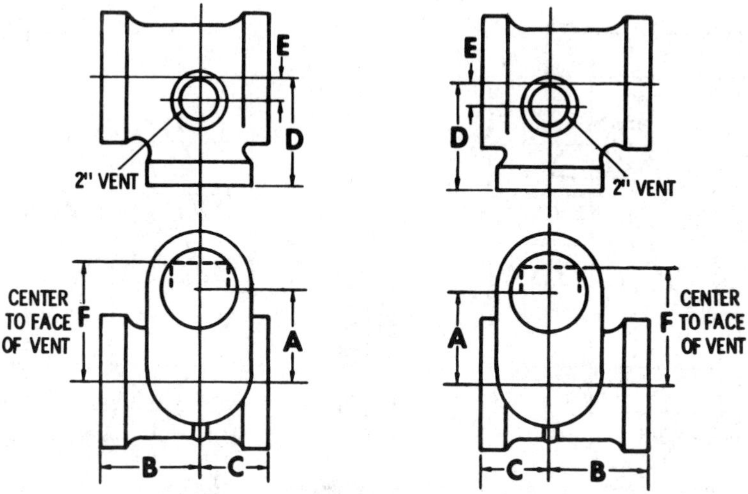

RIGHT-HAND							LEFT-HAND						
4 × 4							5 × 5						
Tapping No.	A In.	B In.	C In.	D In.	E In.	F Center to Face of Vent In.	Tapping No.	A In.	B In.	C In	D In.	E In.	F Center to Face of Vent In.
1	2 5/8	5 1/4	3 1/2	5 1/4	0	3	1	4 7/8	5 3/8	3 3/4	5 3/4	1 1/4	6 1/4
2	2 1/4	5 1/4	3 1/2	5 1/4	0	3	2	4 1/2	5 3/8	3 3/4	5 3/4	1 1/4	6 1/4
3	1 7/8	5 1/4	3 1/2	5 1/4	0	3	3	4 1/8	5 3/8	3 3/4	5 3/4	1 1/4	6 1/4
4	1 1/2	5 1/4	3 1/2	5 1/4	0	3	4	3 3/4	5 3/8	3 3/4	5 3/4	1 1/4	6 1/4
5	1 1/8	5 1/4	3 1/2	5 1/4	0	3	5	3 3/8	5 3/8	3 3/4	5 3/4	1 1/4	6 1/4
6	3/4	5 1/4	3 1/2	5 1/4	0	3	6	3	5 3/8	3 3/4	5 3/4	1 1/4	6 1/4
							7	2 5/8	5 3/8	3 3/4	5 3/4	1 1/4	6 1/4
							8	2 1/4	5 3/8	3 3/4	5 3/4	1 1/4	6 1/4
							9	1 7/8	5 3/8	3 3/4	5 3/4	1 1/4	6 1/4
							10	1 1/2	5 3/8	3 3/4	5 3/4	1 1/4	6 1/4
							11	1 1/8	5 3/8	3 3/4	5 3/4	1 1/4	6 1/4
							12	3/4	5 3/8	3 3/4	5 3/4	1 1/4	6 1/4

CHAPTER 8

Fixtures

As one phase of the construction of a building, the piping and fittings for the plumbing system are roughed-in. As the building nears completion the plumbing fixtures must be installed and connected. Some of the fixtures ordinarily installed are:

Lavatories
Water closets
Bathtubs, with or without showers
Shower stalls
Urinals
Sinks; kitchen, laboratory, mop, etc.
Bidets

LAVATORIES

By definition, a lavatory is *a bowl or basin* used for washing. The term is now applied to the entire unit, which may include a counter top and the appropriate faucets. Lavatories are available in many styles, sizes, and colors to suit nearly any available space and decoration scheme. Most lavatories are designed to be installed in a vanity as shown in Fig. 8-1, and to stand free, as in the pedestal lavatory (Fig. 8-2). Lavatories are made from pressed steel, cast-iron, vitreous china, plastic, or fiberglass.

Courtesy Nutone

Fig. 8-1. A cabinet-top mounted lavatory.

Fig. 8-2. Pedestal-type lavatory.

Supply Connections

There are several methods by which the hot- and cold-water supply can be connected to a faucet. The faucet shown in Fig. 8-3 is made with a threaded shank; there are a number of kinds (Fig. 8-4). In Fig. 8-3 the connections are made with flexible supply tubes.

Some brands of faucets are made with ⅜-in. O.D. flexible copper tubing, extending from the faucet base, instead of threaded shanks. Faucets of this type can be easily connected to the supply piping by using a ⅜-in. O.D. compression union, as shown in Fig. 8-5.

Other faucets require other connections; it depends on the situation. Of all the materials, the easiest to work with is PVC. Typical connections are shown in Fig. 8-6.

Lavatory faucets come in a wide variety of styles from gold-plated ones to those that are in the configuration of animals. The crucial dimension in ordering a replacement faucet is the center-to center measurement, or "centers" (Fig. 8-7). This is the distance between the center of the hot-water handle and the center of the

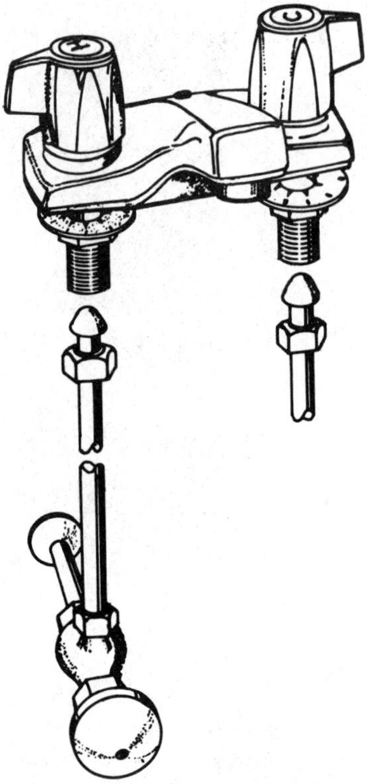

Fig. 8-3. Faucet connections made with flexible supply tubes.

cold-water one. Lavatory faucets commonly have 4-in. centers, but 8 in. is not unusual.

Faucets are usually made of chrome-plated brass, but you can also obtain chrome-plated plastic and plain plastic. Plain plastic is inferior, but coated plastic is good. The chrome-plated brass may be tubular or cast; the latter is superior.

One other kind of faucet is chrome-plated pot metal, which is inferior. To distinguish it from chrome-plated cast brass, just lift one up in each hand. The cast-brass faucet will be heavier.

Waste Connections

Many lavatory bowls having an internal overflow passage use a waste connection with a pop-up drain. Fig. 8-8 shows details of

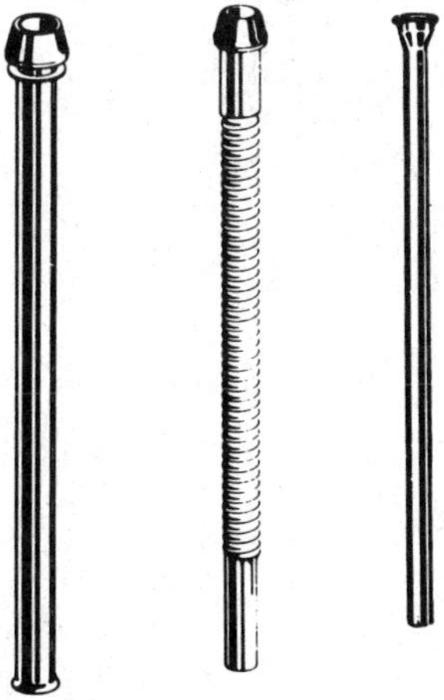

Fig. 8-4. Various types of supply tubes. You can get tubes of chrome-covered copper, corrugated copper, brass (uncoated copper), and PVC plastic.

this type of lavatory drain. In this particular model, the drain stopper is controlled by pushing or pulling a knob located on the faucet assembly. Other pop-up models are available in which a lever is rotated to open and close the stopper. Another type of waste connection is the chain and plug type. Chain and plug drains are rare; most publicly used toilet rooms have lavatories with a plain strainer.

Roughing-In Measurements

The roughing-in measurements are important to the plumber so that he can install supply and drain pipes in the correct locations to connect properly with the lavatory when it is finally set in place. Different styles of lavatories may have different roughing-in dimensions, especially those units from other manufacturers; therefore, it is necessary to obtain all the necessary information before installing the supply and drain lines. This information is usually furnished with the fixture unit, or it can be found in catalogs supplied by the manufacturer. Fig 8-9 shows the roughing-in dimensions for a typical counter-top lavatory.

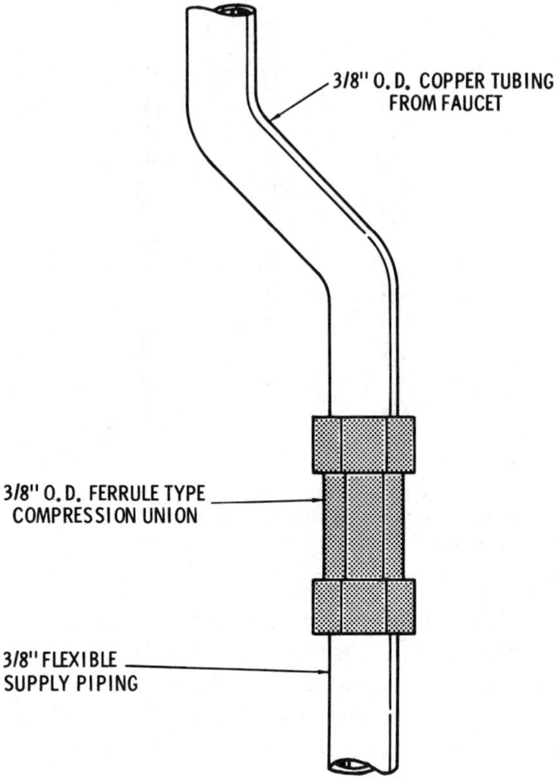

Fig. 8-5. Using a compression union to join $^3/_8$-in. copper tubing from faucet with $^3/_8$-in. flexible supply piping.

Kitchen Sinks

Like lavatories, kitchen sinks (and they are called sinks) are made of different materials. They may be stainless steel, porcelain on steel, or porcelain on cast iron (Fig. 8-10). Cast iron is of higher quality than steel. It should be noted that when looking for stainless-steel sinks that these vary greatly in quality in terms of gauge and finish.

Faucets for sinks may be the compression type, which uses washers, or the so-called washerless kind. These may be either the cartridge or ball type. In the latter, there is a ball with holes in it

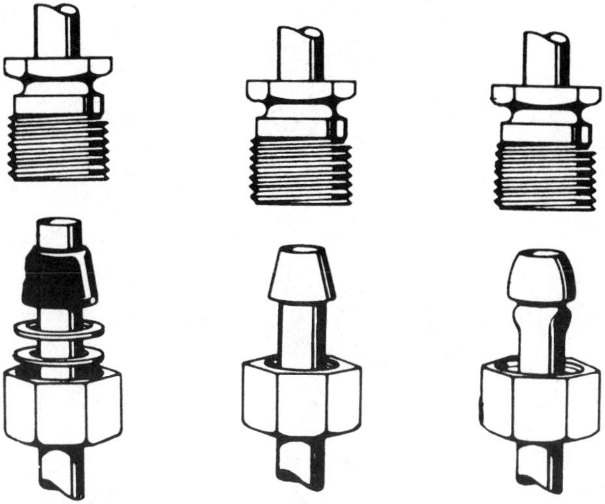

Fig. 8-6. Various ways to connect water-supply tubes to faucets.

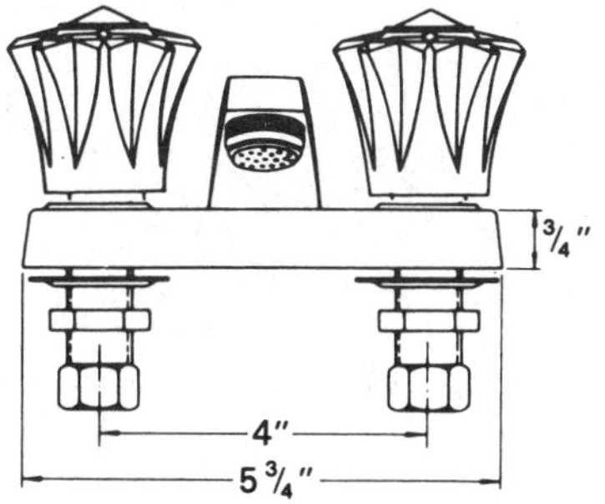

Fig. 8-7. "Centers" on lavatory faucet.

inside the faucet body. When the faucet is open, the holes on the ball are not lined up with those in the body, so water can flow. When closed, the holes are blocked, and the faucet cannot flow.

Faucets may have a pair of handles and a spray hose (Fig. 8-11), but you can also get the single handle kind. Manipulating the single handle enables you to control water flow. Such faucets are always washerless. Washerless faucets last longer than compression types, although repair parts for washerless faucets are normally more costly. Most plumbers favor washerless faucets. The unit is examined more closely later.

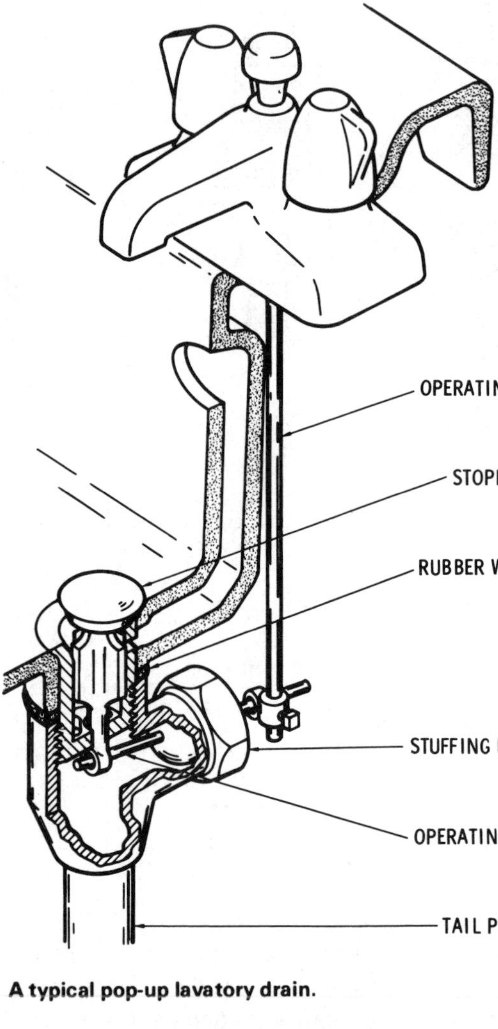

Fig. 8-8. A typical pop-up lavatory drain.

BATHTUBS

Bathtubs are available in various styles, sizes, colors, and materials. The most popular bathtub is the standard oblong-shaped model ranging in length from 4 to 6 ft.; 5 ft. is most common. Square models, such as the tub shown in Fig. 8-12 are available

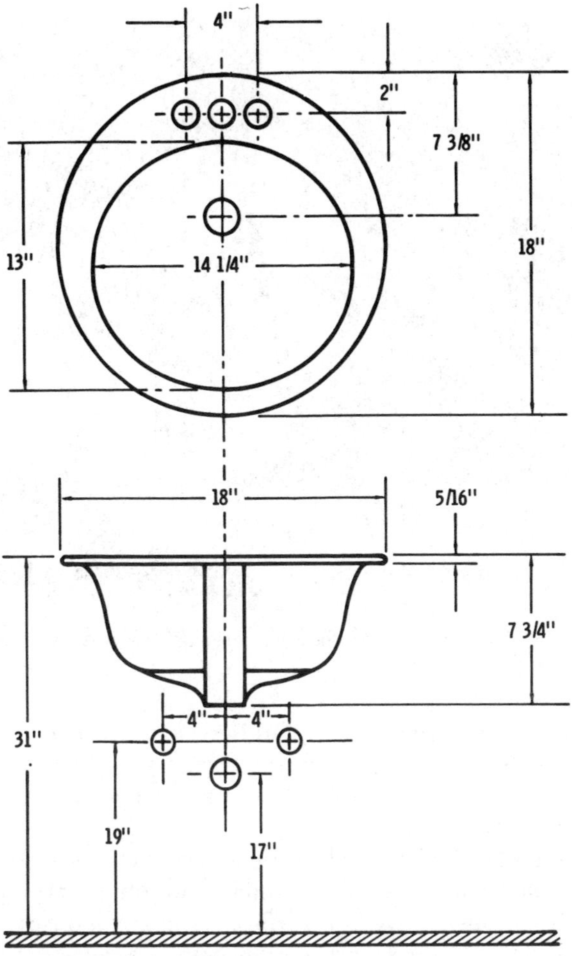

Fig. 8-9. Roughing-in dimensions for counter-top lavatory.

where space is limited or for special installations. Although this type of bathtub is usually referred to as "square," it actually may be slightly longer in one dimension than in the other.

Bathtubs are made of steel covered with porcelain, enameled cast iron, or fiberglass. Cast iron is the heaviest material used, and is more durable than the other types. Cast-iron tubs are, though,

Fig. 8-10. Sinks are made in a variety of materials. This one is enameled cast iron.

more difficult to install. Many plumbers favor steel because of its lighter weight, but steel bathtubs can be noisy, and some kinds are made in two parts with a welded seam that shows, a detail people do not like.

The fiberglass bathtub, which can come in one piece or as part of a kit consisting of the tub surround and tub, is very light and relatively simple to install. Its surface is smooth and easy to keep clean, but it does scratch fairly readily; kits are available to remove the scratches. See Figs. 8-13, 8-14, and 8-15.

You can also obtain a one-piece tub-shower unit. Its seamlessness is a definite advantage in maintenance and cleaning. You can get fiberglass tub-shower units in a wide variety of shapes. The standard 5-ft. tub is the most common.

Several varieties of traps are available for bathtub installations. A check should be made with local authorities to determine which types are approved for a given locality.

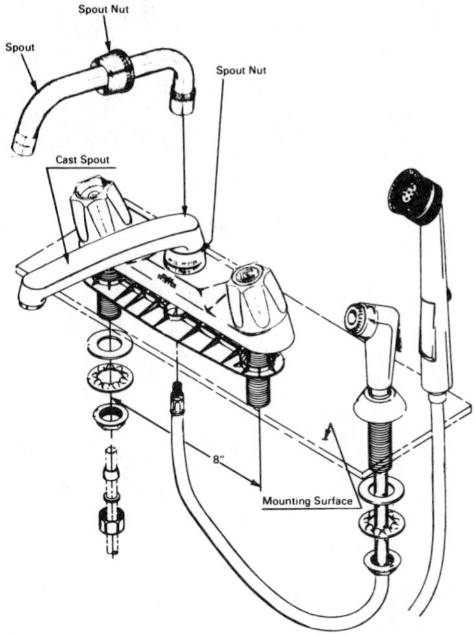

Fig. 8-11. Exploded view of a kitchen faucet.

Fig. 8-12. A tub that is called a square tub, though it is not actually square.

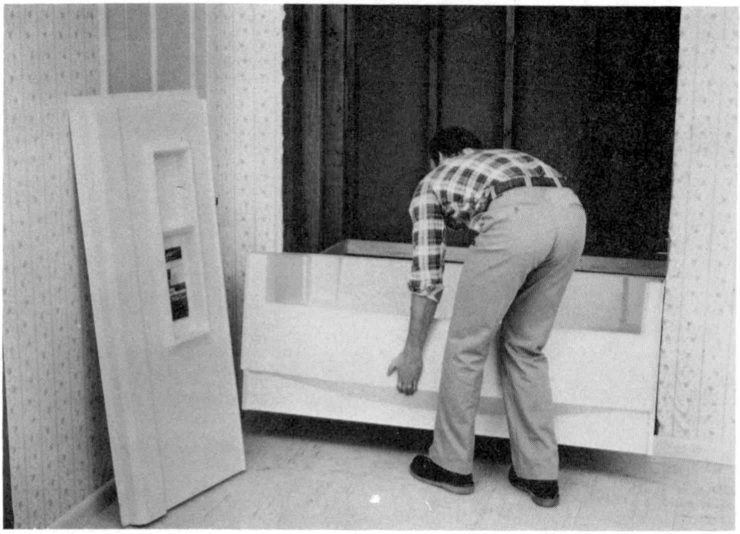

Courtesy Owens Corning

Fig. 8-13. A fiberglass tub is easy to handle.

Fig. 8-14. Plumbing is hooked on to drilled holes.

Fig. 8-15. The resulting unit is easy to keep clean and comes in a variety of colors.

Fittings

A variety of fittings are used on tubs. What follows is a summing up.

Tub and shower faucets come with three-valve diverters, two-valve diverters, two-valve shower fittings, or two-valve tub fillers and single-control tub and shower units (Fig. 8-16).

A three-valve diverter has hot- and cold-water faucets, called valves, and a faucet valve in the middle for diverting the water from the shower to the bath. The two-valve diverter has hot- and cold-water faucets and a pop-up device on the spout for diverting the water as needed. The two-valve shower fitting has hot- and cold-water faucets for filling the tub. The single-control unit, as its name implies, has one handle.

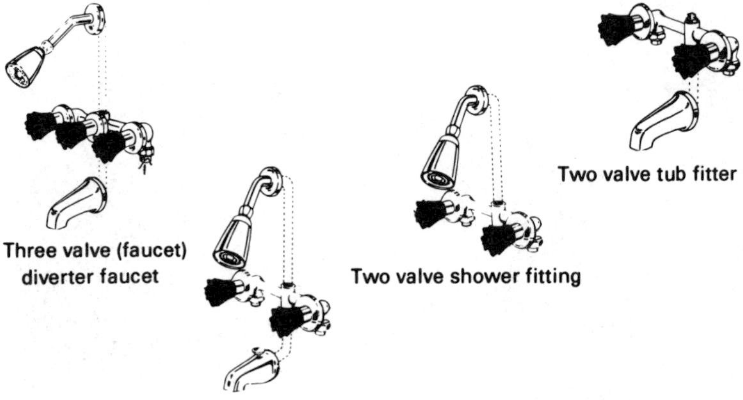

Fig. 8-16. Various tub valve arrangements.

The standard centers on tub faucets is 8 in., but 6 in. and 11 in. are not uncommon when the fittings are wall mounted. On an old-fashioned freestanding tub, the measurement is only 3½ in.

Faucets for tubs are available in the same materials as those for lavatories, and the same quality considerations apply.

The compression-type faucet is most commonly used on tubs, and should repairs be required, they will be difficult, mainly

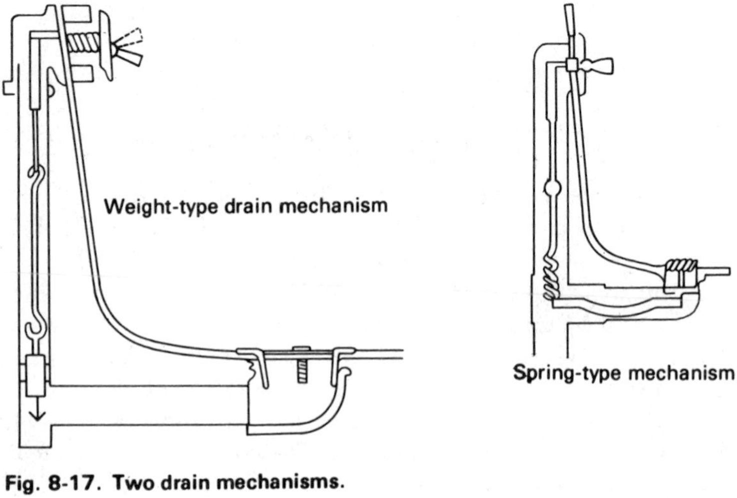

Fig. 8-17. Two drain mechanisms.

because the mechanisms are inside a wall and must be taken out with the help of a socket wrench.

Shower heads today are threaded on both ends. To replace the head on the shower, you just unscrew it. The older shower head required unscrewing the head and the shower arm.

Drain Hardware

Two types of drain mechanisms are normally used on tubs: the weight and the spring (Fig. 8-17).

The weight type consists of a lever that sticks through an escutcheon on the tub. Linked to the lever, hanging down inside a drain pipe, is a weight. Flip the lever up, and the weight lowers into a seat, blocking water; push the lever down, and the weight lifts off the seat, letting water flow.

The spring mechanism is present if a tub has a chrome pop-up stopper. Lift the lever and a linkage with a spring on the end presses against a rocker-arm assembly that opens the stopper—pushes it up. Push the lever down, and the stopper closes.

Roughing-In Measurements

The roughing-in measurements for bathtubs are supplied with the unit, or they can be found in the manufacturer's catalog. Care must be taken to position the supply and drain lines correctly and to place any necessary hanger hardware in the proper position . Fig. 8-18 lists the roughing-in dimensions for a typical rectangular built-in bathtub. The dimensions for a typical square bathtub are shown in Fig. 8-19. It should be noted that all bathtubs are fur-

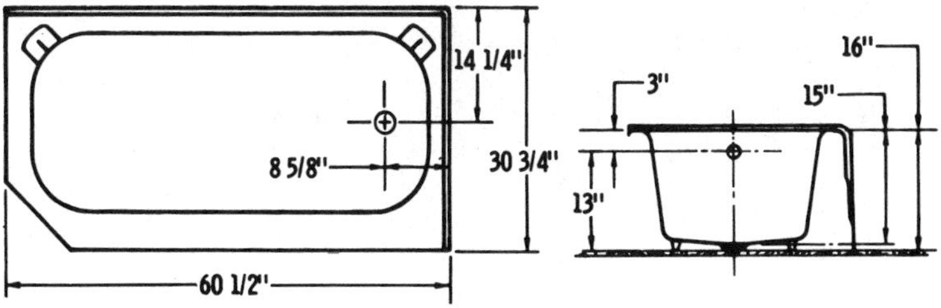

Fig. 8-18. Roughing-in dimensions for a rectangular built-in bathtub installed with finished walls on three sides.

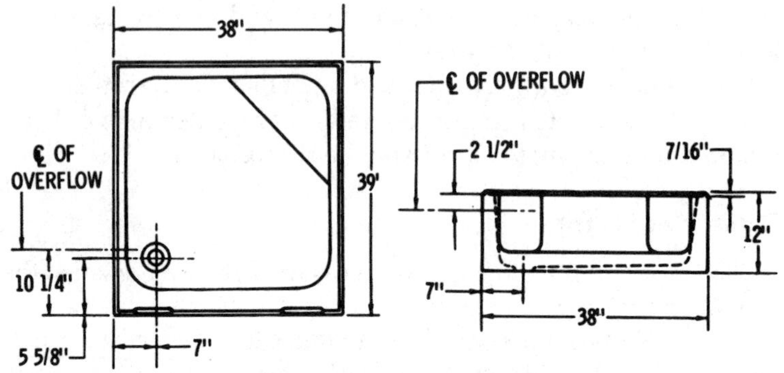

Fig. 8-19. Roughing-in dimensions for a square bathtub having an integral corner and opposite the outlet.

nished with either a right-hand or a left-hand outlet to permit installation along any bathroom wall.

The temperature of the water for the gang-type shower baths installed in many schools and factories can be controlled automatically. One of these control devices is shown in Fig. 8-20. The temperature of the water can be thermostatically controlled for hospital hydrotherapy, film processing, plastic molding, and many industrial processes. The control mixes the hot and cold water and delivers the blended mixture at the desired thermostatically controlled temperature. A liquid-filled expansion-type thermostat, contained within the assembly, maintains the selected water temperature even though both hot- and cold-water supplies may fluctuate in temperature and pressure.

Courtesy The Powers Regulator Co.

Fig. 8-20. Control device used to deliver thermostatically controlled water to the gang-type showers often found in schools and factories.

In the control device shown (Fig. 8-20), the water flow from the mixer is stopped automatically and immediately in event of a failure in the supply of cold water. A concealed adjustment for raising and lowering the temperature range is included. A single handle selects the temperature of the water delivered, and it also shuts off the discharge of water. The control device is constructed with only one moving part and is accessible from the front. It is available in two different temperature ranges—65°F to 110°F and 75°F to 175°F.

A dial-type thermometer (Fig. 8-21) is available either in rigid or remote bulb-type instruments that are placed in a pipe line for immersion in a liquid or for air exposure. The instrument is available in a wide variety of temperature ranges. This type of thermometer permits a visible check on the operating temperature of the water for a gang-type shower bath.

The control fixture and the dial-type thermometer can be enclosed in a cabinet, with only authorized personnel having

Fig. 8-21. A dial thermometer permits a visible check on the operating temperature of the water for a gang-type shower.

access to the cabinet (Fig. 8-22). The dial-type thermometer permits a visible check on the operating temperature of the water for the shower bath at any time.

TOILETS

The careful selection and installation of the toilet or water closet is more important than with any other fixture.

Toilet Bowls

Four different general types of toilet bowls are available. They are: (1) siphon-jet; (2) blow-out; (3) reverse-trap; (4) washdown.

Siphon-Jet—The flushing action in the siphon-jet bowl (Fig. 8-23) is accomplished by directing a jet of water through the upward leg of the trapway, which fills the trapway and starts the siphoning action at the same time. The strong, quick, and relatively quiet action of the siphon-jet bowl, together with its deep

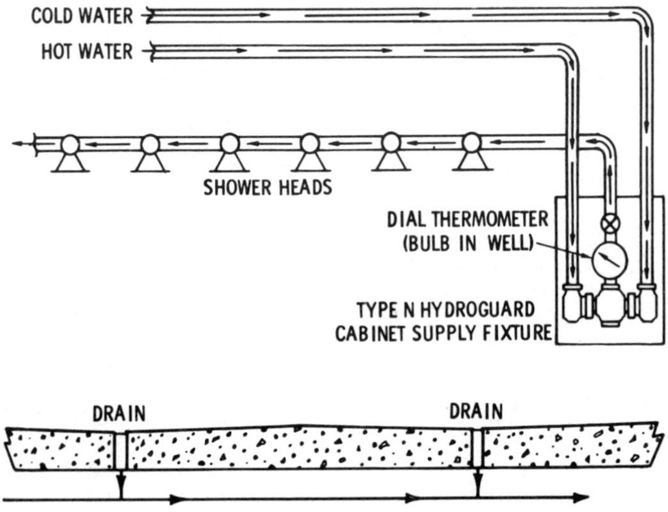

Courtesy The Powers Regulator Co.

Fig. 8-22. The control fixture and dial thermometer can be enclosed in a cabinet. Only authorized personnel should be permitted access.

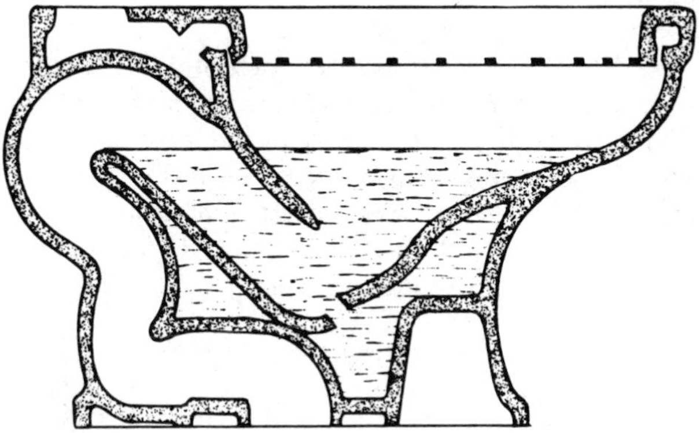

Fig. 8-23. A siphon-jet toilet bowl.

water seal and large water surface, is recongized universally by sanitation authorities as the most satisfactory toilet bowl in existence.

Blow-out—Since the blow-out type of toilet bowl (Fig. 8-24) depends entirely on a driving jet action for its efficiency, rather than on siphoning action in the trapway, it cannot be compared favorably with other types of toilet bowls. It is economical in its use of water; however, it does have a large water surface that reduces fouling space, a deep water seal, and a large unrestricted trapway. The blow-out bowls are well suited for use in schools,

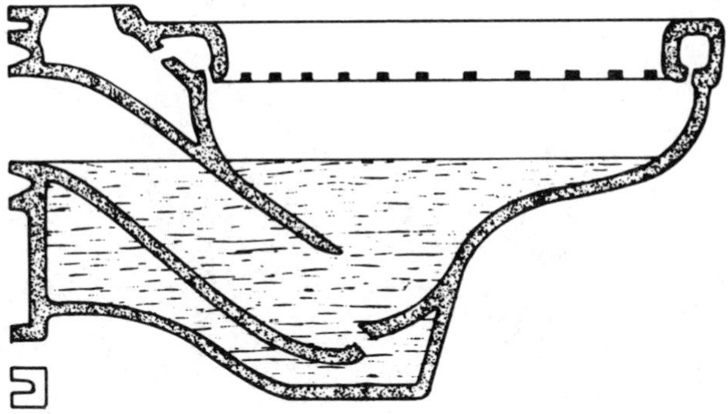

Fig. 8-24. Blow-out toilet bowl.

offices, and public buildings. They are operated only by flush valves.

Reverse-Trap—The general appearance and flushing action of the reverse-trap toilet bowl (Fig. 8-25) are similar to the siphon-jet bowl. The water surface and the size of the trapway are smaller, and the depth of the water is less; therefore, less water is required for operation. The reverse-trap bowls usually are suitable for installation with either a flush valve or a low supply tank.

Washdown—The washdown-type bowl (Fig. 8-26) is simple in

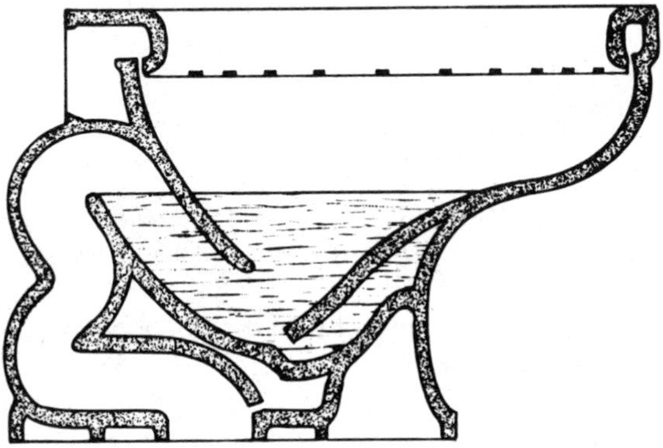

Fig. 8-25. Reverse-trap toilet.

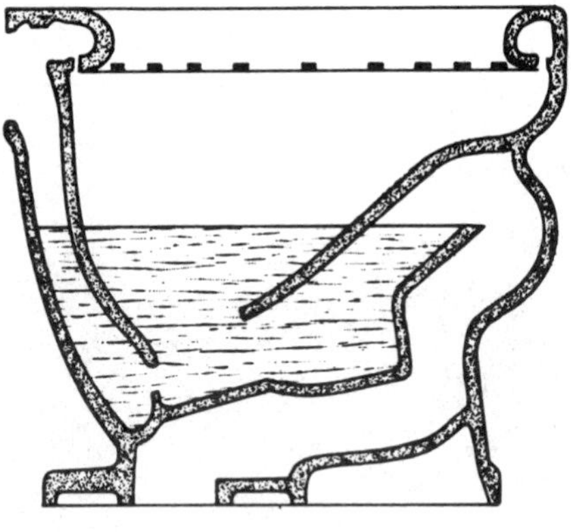

Fig. 8-26. A washdown toilet bowl.

construction and is highly efficient within its limitations. Proper functioning of the bowl is dependent on siphoning action in the trapway, accelerated by the force of water from the jet directed over the dam. Washdown bowls are used widely where low cost is a major factor, but they have large areas of exposed china and are subject to fouling. Hence, many municipalities have discontinued their use.

The wall-hung tank toilet has been largely supplanted by the freestanding (tank bolted to the bowl) type shown in Fig. 8-27.

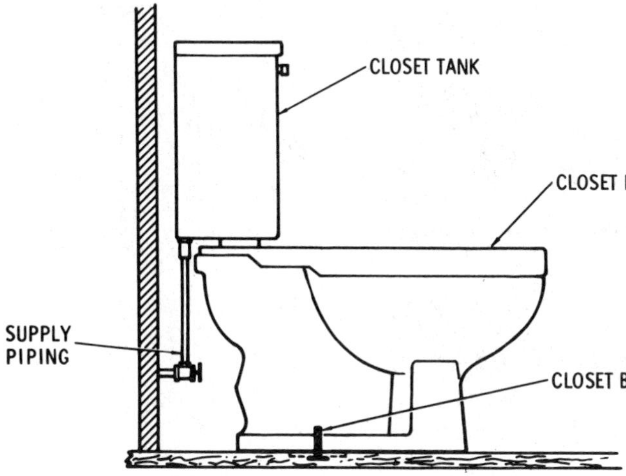

Fig. 8-27. Freestanding toilet.

Stoppages in toilet bowls usually are a result of foreign objects falling into the bowl. Normally, these obstructions can be cleared by either a force cup or an auger. If neither the force cup nor the auger can remove the obstruction, removal of the toilet bowl may be necessary. To remove the toilet bowl, refer to Figs. 8-28 and 8-29 and proceed as follows.

1. Shut off the water supply and empty the flush tank by siphoning or sponging.
2. Disconnect the water pipe from the tank.
3. Remove the tank from the bowl if it is a two-piece unit; or

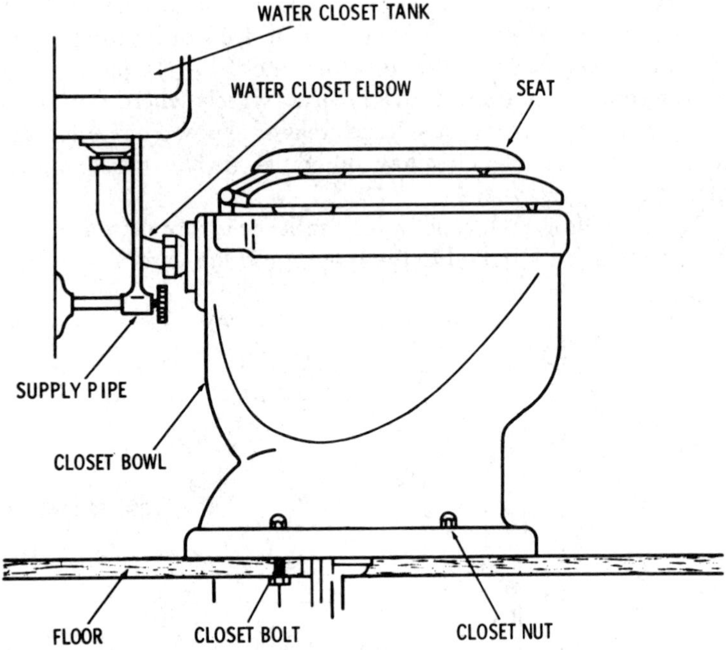

Fig. 8-28. Toilet bowl details.

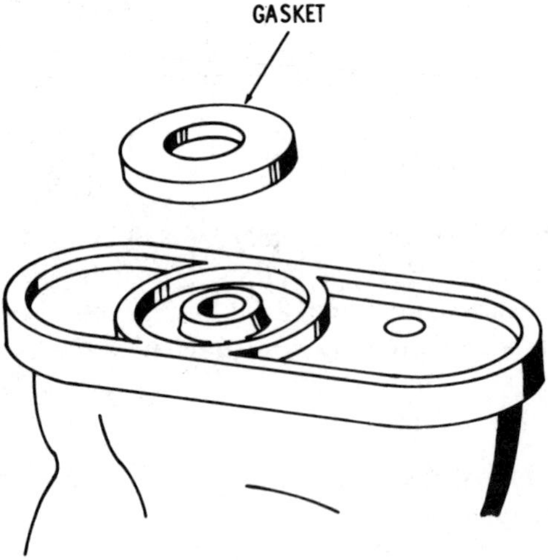

Fig. 8-29. Bottom view of toilet bowl base.

disconnect the tank-bowl pipe connection if it is a wall-hung unit.

4. Remove the bowl seat and cover.
5. Remove the bolt covers from the base of the bowl and remove the bolts holding the bowl to the floor.
6. Break the seal at the bottom by jarring the bowl, and lift the bowl free.
7. Remove the obstruction from the discharge section.

To reinstall the toilet bowl, proceed as follows:

1. Obtain a wax seal or gasket from a plumbing supply dealer.
2. Clean the bottom portion of the bowl, place the wax seal or gasket around the bowl horn, and press it into place.
3. Set the bowl above the soil pipe, and press it into place.
4. Install the floor flange bolts; draw them snug, but do not overtighten them. Overtightening may crack the base of the bowl. A level should be used while tightening the bolts to make sure that the bowl is level, using shims if necessary.
5. Reinstall the items that were removed including bolt covers, tank and water pipe connections, seat, and cover.

The connection of the toilet bowl to the soil pipe is an important part of the toilet installation. Fig 8-30 shows a typical installation using a lead bend soldered to a brass floor flange. The brass floor flange is shown in Figs. 8-31 and 8-32.

The brass floor flange must be secured to the floor when connecting to a lead pipe. As shown in Fig. 8-30 and 8-33 screws may be used when the flange is fastened to a wood floor; bolts should be used for fastening to concrete. A brass flange may also be soldered to a copper pipe as shown in Fig. 8-34.

As shown in Fig. 8-35, it is not necessary to secure the cast-iron flange to the floor when connecting it to cast-iron soilpipe. Fig. 8-36 illustrates how a DWV plastic pipe and a DWV plastic closet flange may also be used for the toilet drain connection.

Toilet Tanks

Tanks are used to supply water for flushing in residences or other buildings where quiet flushing is desired or necessary. A

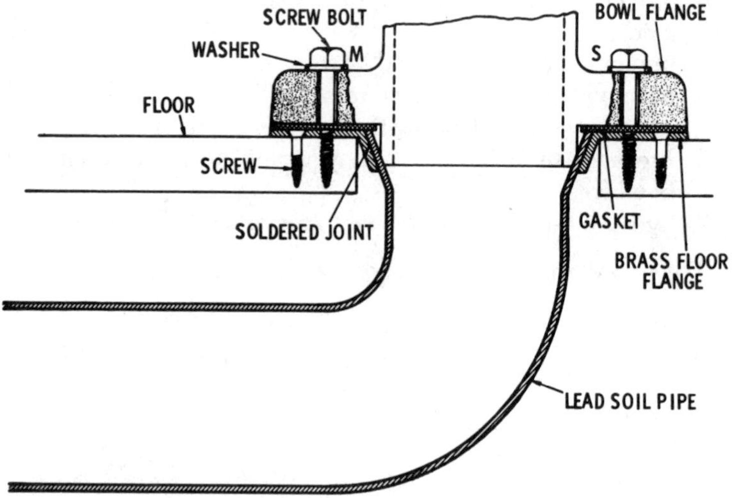

Fig. 8-30. Lead-pipe elbow used to connect toilet bowl with soil pipe.

minimum quantity of water is used during the flushing process. If noise is not a factor, diaphragm-type flushing valves are used. Although tank designs may vary, their mechanisms are similar.

The most common type of flush mechanism is the ballcock, which is shown is Fig. 8-37. Over the last few years, however, mechanisms that are plastic units which are quieter than the standard ballcock and less expensive have been installed, though not without occasional problems; the plastic unit does work quite well (Fig. 8-38).

Toilets installed in residences are usually tank toilets. Both tank and flush valve water closets are installed in schools, commercial,

Fig. 8-31. Typical cast-brass closet floor flange for lead pipe.

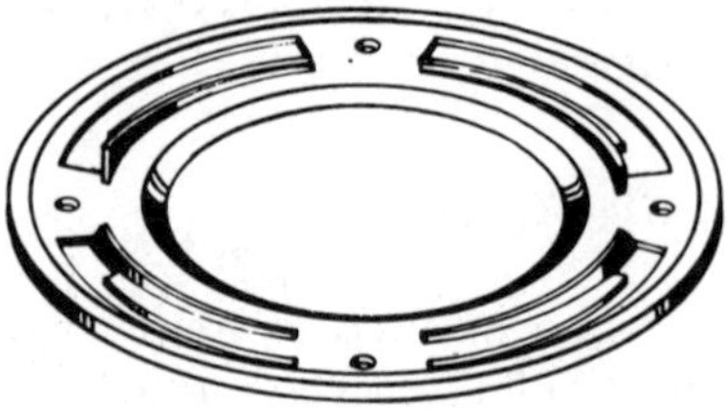

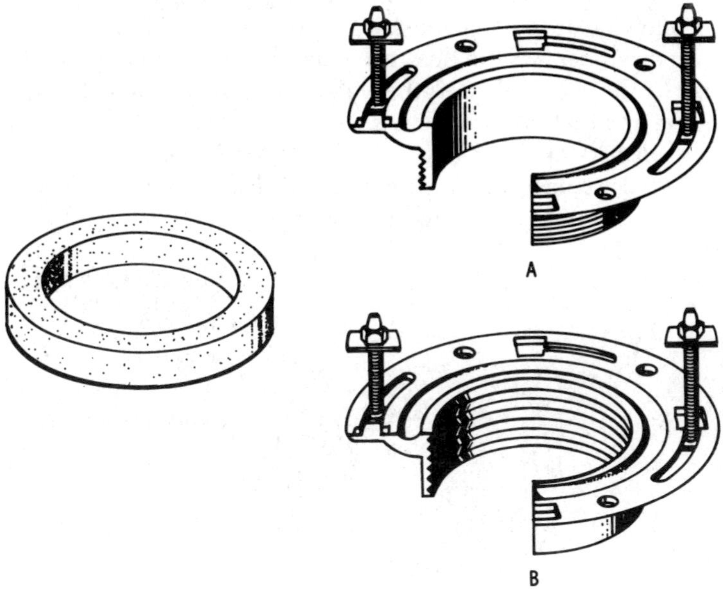

Fig. 8-32. Brass closet flanges: (A) male and (B) female with wax setting ring, bolts, and nuts for connecting 4-in. steel pipe.

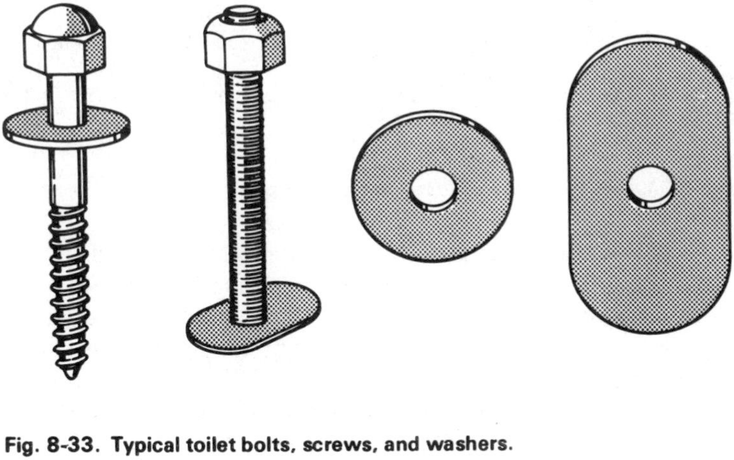

Fig. 8-33. Typical toilet bolts, screws, and washers.

and industrial buildings. The unit can be glued in place.

Following are some common problems and solutions.

Float Valves—If the float valve leaks, a worn plunger washer may be the cause. Shut off the water and drain the tank. Remove the screws that hold the levers, remove the plunger, and install new washers on the plunger. If the ballcock and float ball are corroded, replace them; preferably with one of the newer fill valves that need no float rod or float ball.

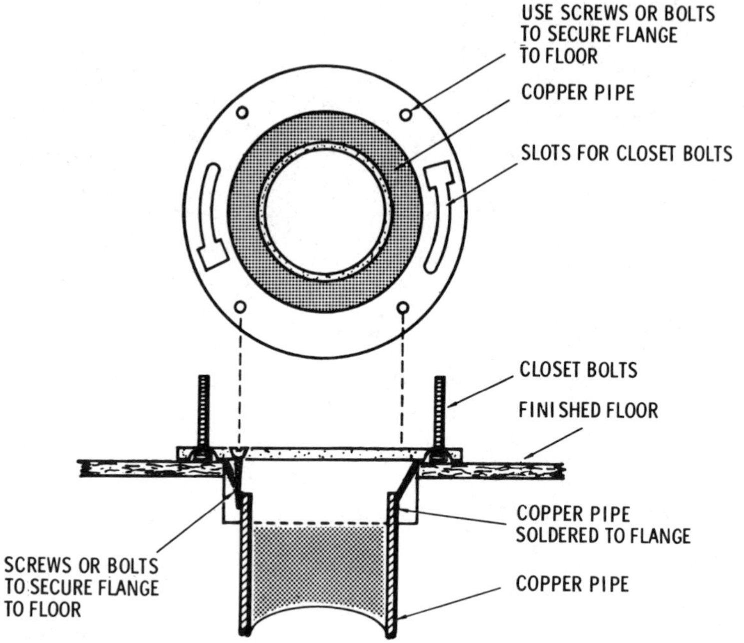

Fig. 8-34. A brass flange soldered to a copper pipe.

Flush Valves—Occasionally, the rubber ball of the flush valve becomes soft and loses its shape, resulting in poor seating. If this occurs, replace the ball and at the same time remove all the corrosion from the lift wire to prevent binding of the wire in the travel guide. It is also possible to get a new flush seat if this is the problem. Many state and city codes now specify elongated closets, as seen in Fig. 8-40, for public use.

Float—The float maintains the level of the water in the flush tank at approximately 1 in. below the top of the overflow tube; this provides sufficient water for proper flushing action. If a leak developes in the ball-type float, it does not rise to the correct height, and the intake valve remains open, causing a continuous flow of water. The float should be replaced in such a case.

Tank Condensation—Moisture condensation on the outside surface of the flush tank can be avoided by installing an insulating jacket, which can be purchased at most plumbing stores. Also, a

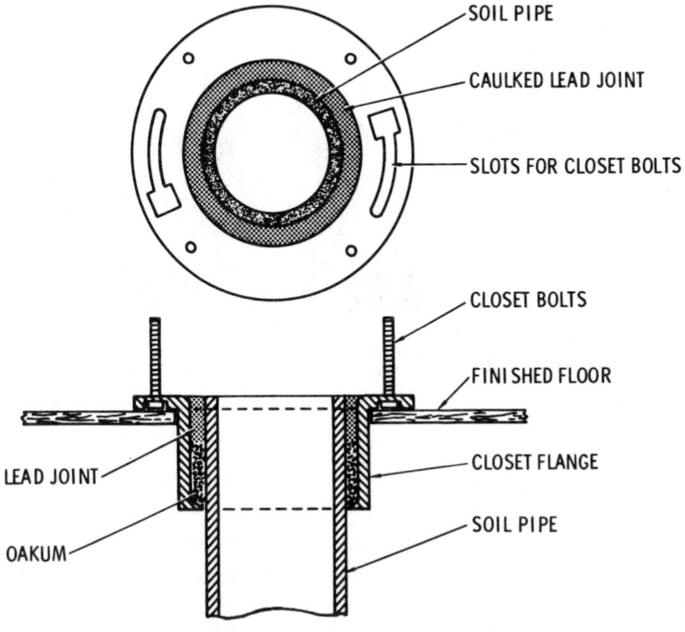

Fig. 8-35. Cast-iron flange connected to cast-iron soil pipe using a lead and oakum joint.

device that mixes some hot water with the cold water entering the tank is available. Either device prevents condensation by keeping the temperature of the exterior surface of the tank higher than the dew point of the surrounding air.

Roughing-in dimensions for a vitreous china close-coupled combination toilet tank and bowl with siphon-jet, whirlpool

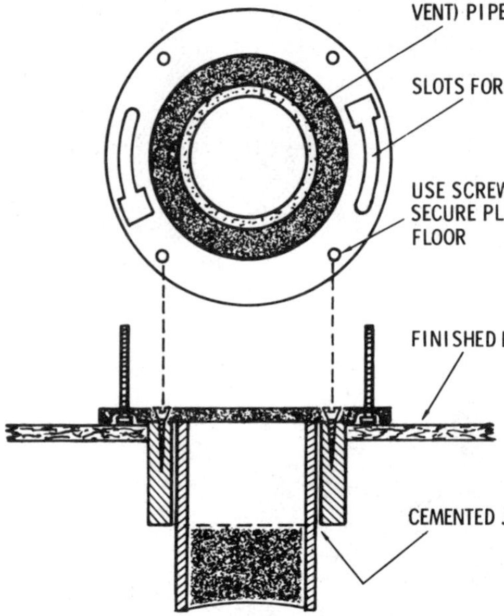

Fig. 8-36. Using DWV plastic pipe and a DWV plastic closet flange.

action, and an elongated rim are diagrammed in Fig. 8-40. The dimensions may vary by ½ in.; installation should be made in accordance with the on-site conditions. The roughing-in dimensions for a vitreous china close-coupled combination toilet tank and bowl with reverse-trap, whirlpool action, and regular rim are diagrammed in Fig. 8-41.

Several methods are used to make the water supply or flush pipe connections. The portion of the toilet bowl that receives the connection is called the inlet horn, and it may be located on the end, top, or side. The flush connection may be exposed with a foot-type flush valve (Fig. 8-42). A foot-pedal flush valve is often used for institutional toilets.

Concealed flush valves with oscillating handles and different types of elbow flush connections are shown in Fig. 8-43. A vandal-proof push-buttom flush valve is also shown.

A vitreous china wall-hung toilet bowl with twin self-draining jets and a flush valve with a vacuum breaker are diagrammed in Fig. 8-44. Concealed supporting-chair units with drainage fittings for siphon-jet and blow-out type toilet bowls are shown in Fig. 8-45A and B. Concealed supporting-flange units for blow-out and siphon-jet bowls are shown in Fig. 8-45C and D.

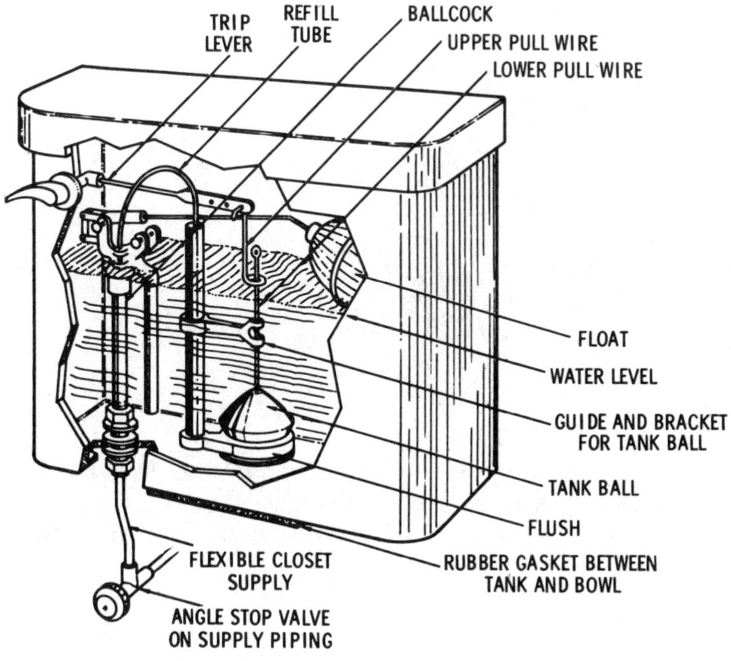

Fig. 8-37. Flush-tank mechanism for a toilet bowl.

URINALS

Since urinals are commonly installed in public restrooms and are subject to hard usage, it is essential that they possess features that enable them to be kept clean and free of debris as possible. Frequently, the toilet is used (rather than the urinal), which may be an entirely sanitary practice; however, the toilet is too low for convenient use as a urinal.

Urinals should be constructed from a material that is nonabsorbent and noncorrosive. Wood should never be used because it is

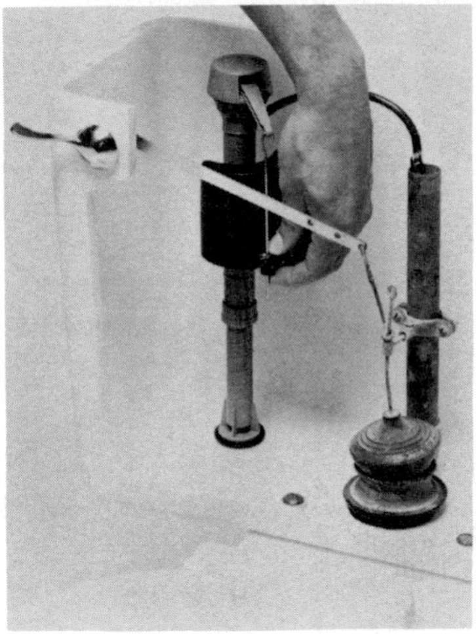

Fig. 8-38. One of newer ballcock mechanisms.

absorbent, and iron is corrosive. The two materials best suited for this purpose are earthenware and vitreous china.

Several different types of urinals are in use. These kinds are:

1. Individual wall-type.
2. Pedestal.
3. Stall.

Individual Wall

Several different types of individual wall urinals are used. The wall unit usually consists of a bowl attached to the wall at a convenient height and a means for flushing and discharging the waste. Two general shapes (round and lipped) of bowls are commonly used (Fig. 8-46); the lipped type is more desirable.

The two common methods of flushing used are: (1) the washdown type; and (2) the siphon-jet type. The lip-type urinals should

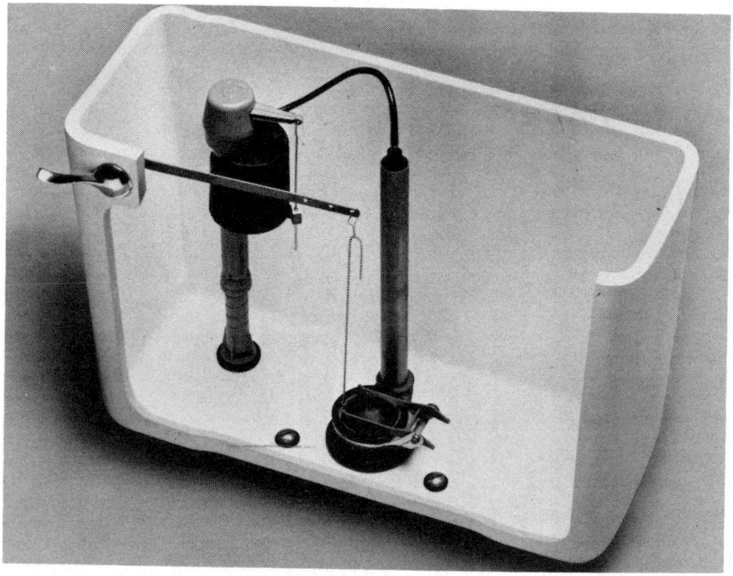

Fig. 8-39. If necessary, a new seat can be glued in place over the old, damaged one.

be of the flushing-rim type. This permits thorough cleansing of the entire interior surface at each flushing.

Since the bowl of the washdown type of urinal does not carry a standing body of water, an offensive odor may result unless the urinal is flushed each time that it is used (Fig. 8-47). This is not a disadvantage in the siphon-jet urinal because a quantity of water remains in the bowl after each flushing.

Urinals are made, preferably, from vitreous china. This is a clay material that is fired to a high temperature and becomes extremely hard. After the first firing, the hard vitrified body is covered with a glaze and fired again. It should be kept in mind that this firing occurs at such high temperature that the piece reaches a molten stage, and the hard body and glaze become a single homogeneous mass. Vitreous china does not craze, crack, or discolor under severe usage. Acids are not injurious to vitreous china. The surface is actually a part of the body, which is hard, nonporous, and impervious to moisture—as the name *vitreous* implies. Also, blows

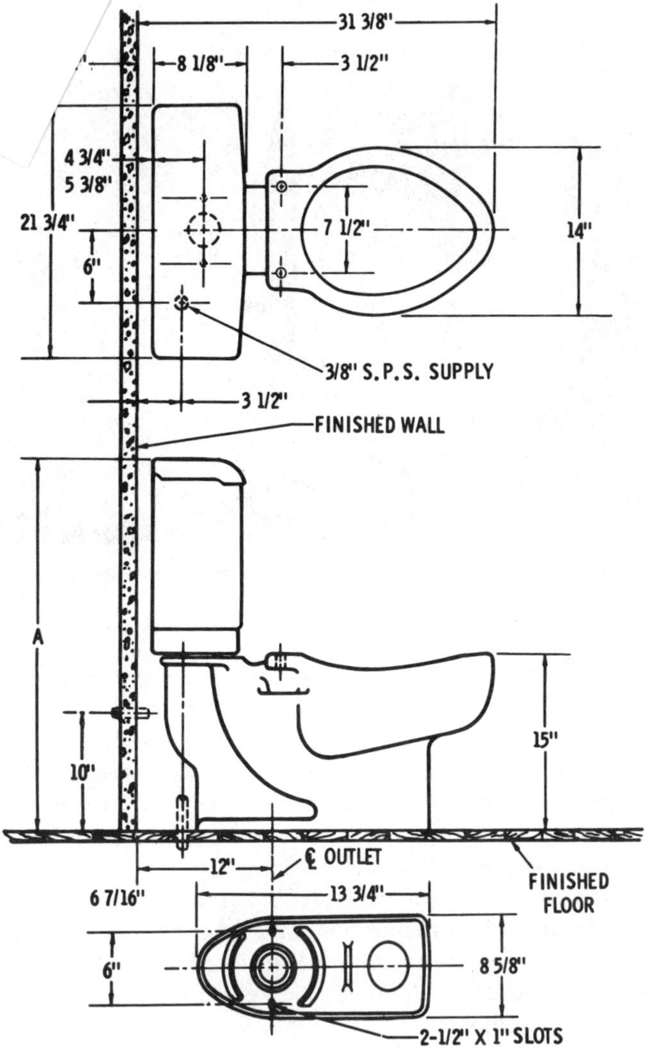

Courtesy Crane Co.

Fig. 8-40. Roughing-in dimensions for a vitreous china close-coupled combination toilet tank and bowl with siphon-jet, whirlpool action, and an elongated rim.

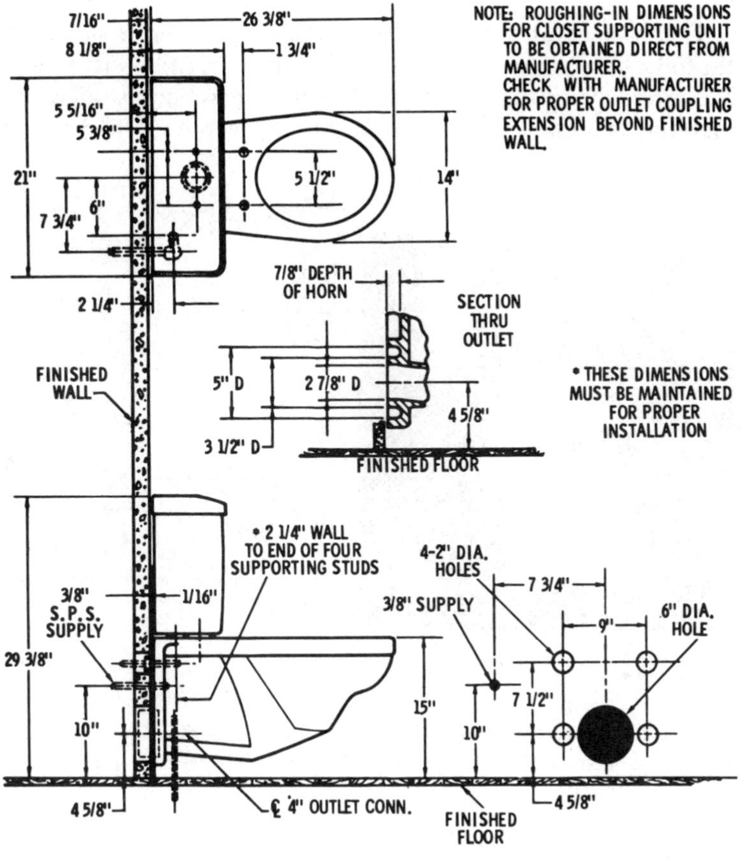

Courtesy Crane Co.

Fig. 8-41. Roughing-in dimensions for a vitreous china close-coupled combination toilet tank and bowl with reverse-trap whirlpool action and a regular rim.

received from falling cups, tumblers, and bottles, which sometimes break lavatories made from common earthenware, do not damage vitreous china. Since the body and glaze cannot be separated on vitreous china, there is no enamel or other surface to loosen, chip, or peel off.

Pedestal

The basic construction of the pedestal-type urinal is similar to the toilet bowl, since it is flushed and cleaned by siphon-jet action. It is probably the most sanitary of all the urinals because all waste material is removed with each flushing action. The passage

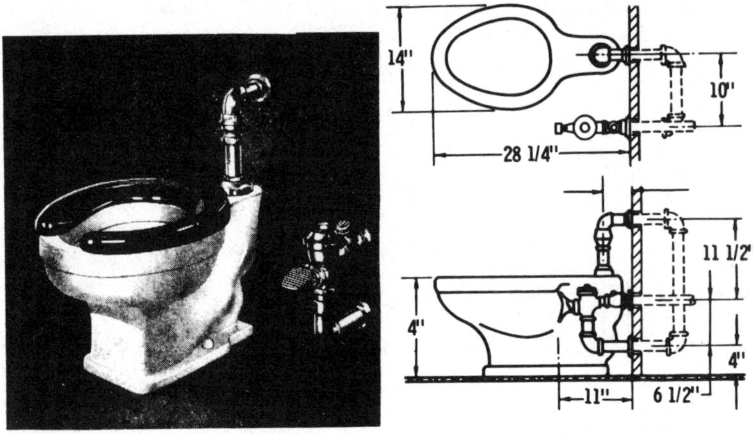

Courtesy Crane Co.

Fig. 8-42. The flush connection may be exposed with a foot-type flush valve. Foot-pedal flush valves are often used for institutional toilets.

through the trap is just as large as that for most toilets, and because of the siphon-jet action, clogging is almost impossible. Installation dimensions for a typical siphon-jet action pedestal urinal are shown in Fig. 8-48.

Stall

These urinals are usually made of vitreous china and are designed with careful consideration of the hygienic principles essential to good health. The sloped front design encourages closer approaches over the lip of the urinal; consequently, improvement in restroom cleanliness and easier maintenance result (Fig. 8-49).

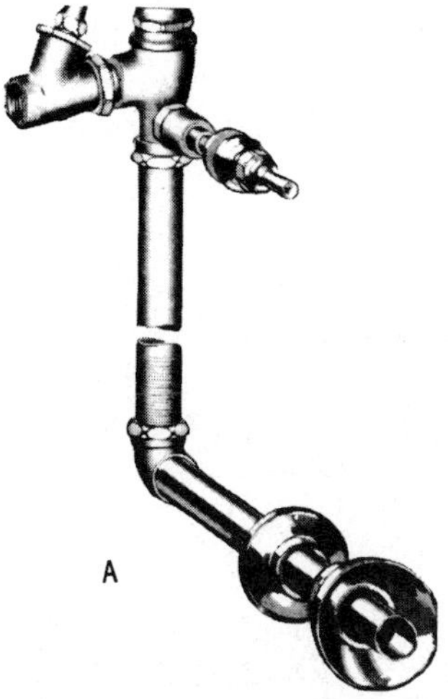

(A) Semiconcealed elbow flush connection.

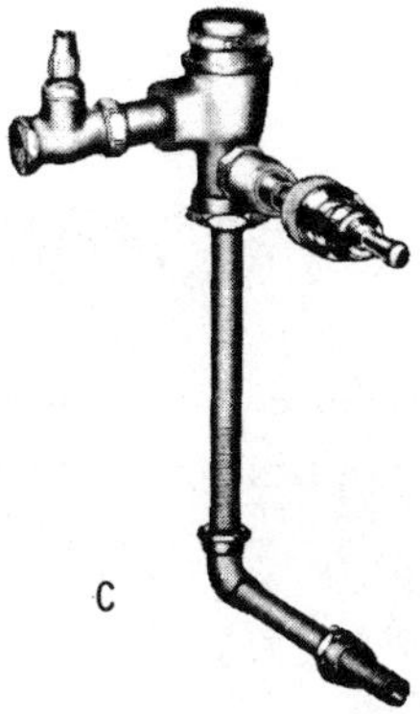

(C) Concealed elbow flush connection.

(B) Concealed double-elbow flush connection.

(D) Vandal-proof push-button-type flush valve.

Courtesy Crane Co.

Fig. 8-43. Concealed flush valves with oscillating handles and various types of elbow connections.

The drip receptor is unusually large, and the sides are straight, facilitating tile setting. These urinals are available with integral flushing rims that distribute the flushing action evenly to cleanse the stall interiors thoroughly.

Installation dimensions for a battery of two stall urinals with the flush tank and seam are shown in Fig. 8-50. Since the urinals are made of vitreous china, careful handling and a knowledge of the correct installation procedure are imperative. Damage to the stalls

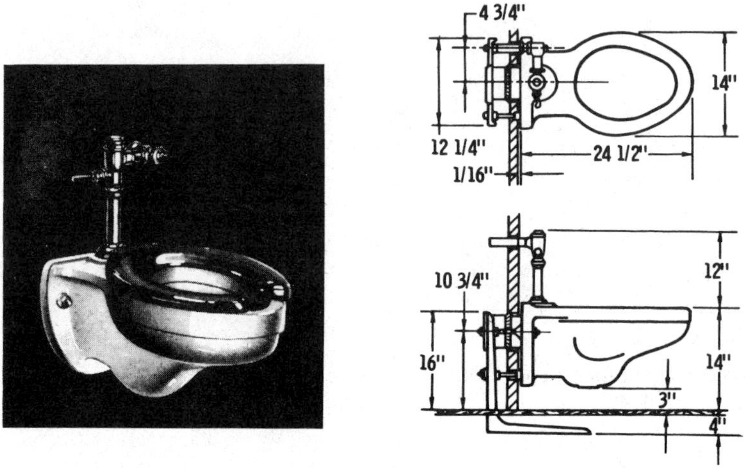

Courtesy Crane Co.

Fig. 8-44. Installation dimensions for a vitreous china wall-hung toilet bowl with twin self-draining jets and flush-valve vacuum breaker.

resulting from unnecessary strains can be reduced greatly, or avoided entirely, if the following suggestions are adhered to (see Fig. 8-51):

1. A pit (see Fig. 8-51A) should be provided in the rough floor; sufficient depth should be allotted for the lip of the urinal to set flush with the finished floor, with a sand cushion of at least 1 in. underneath the stall.
2. A space of at least ½ in. should be provided between the edge of the pit in the rough floor and the stall (see Fig. 8-51B). The waterproof strip, which is furnished with the urinal, should be placed around the urinal.
3. A space of not less than ⅛ inch should be provided between

Concealed supporting-chair unit for:

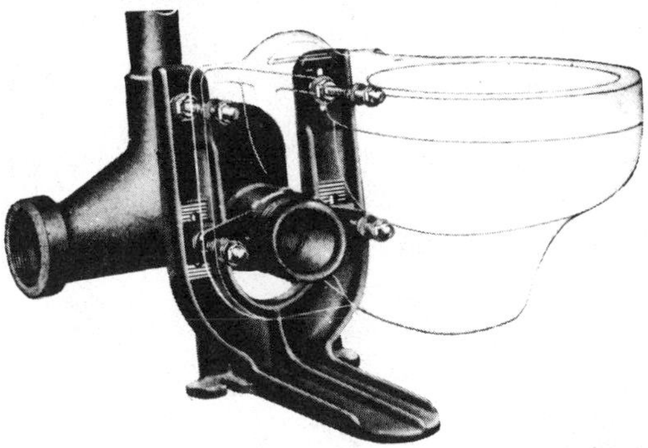

(A) Siphon-jet bowl

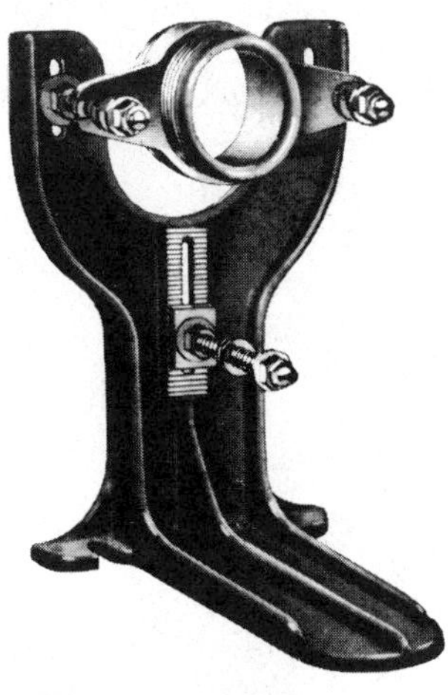

(B) Blow-out-type bowl

Concealed supporting-flange unit for:

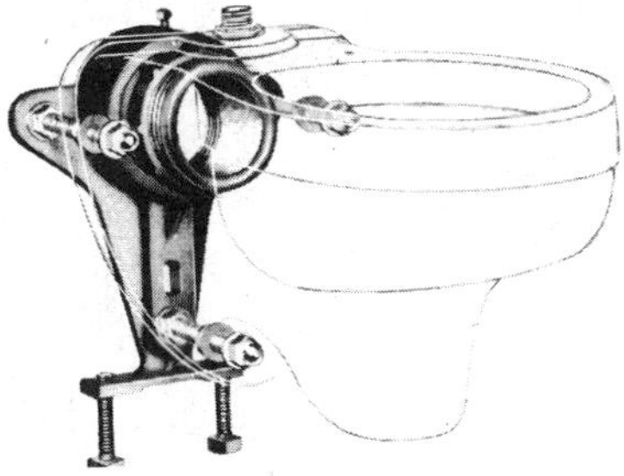

(C) Blow-out-type bowl

(D) Siphon-jet bowl

Courtesy Crane Co.

Fig. 8-45. Concealed supporting chair and supporting flange units.

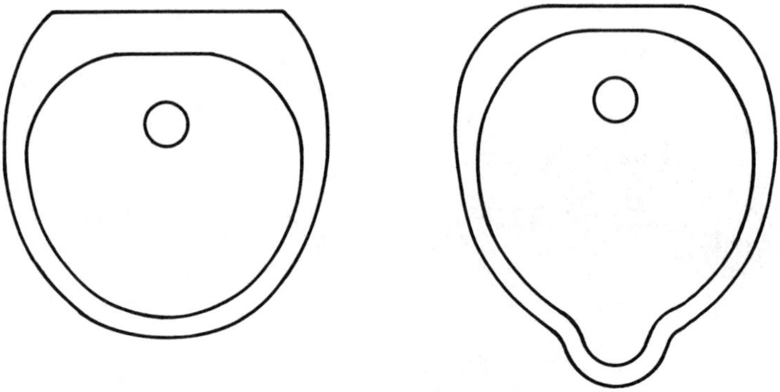

Fig. 8-46. Top view of stall urinal.

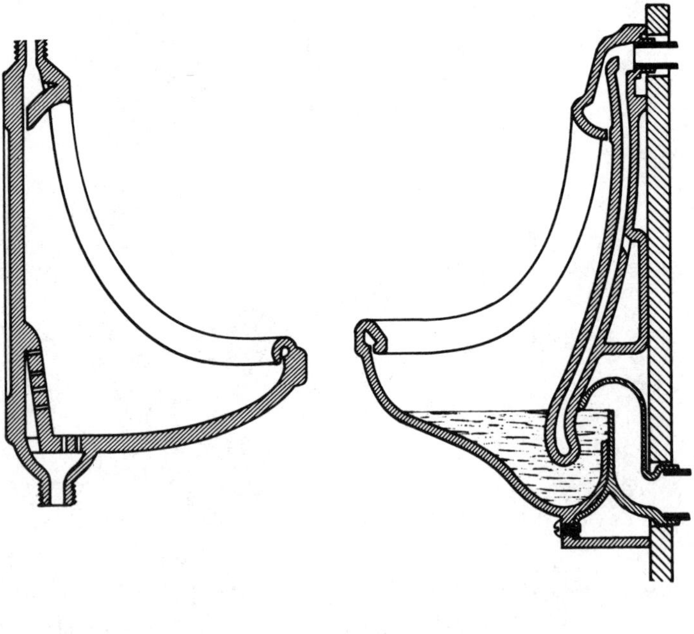

Fig. 8-47. The washdown (left) and siphon-jet (right) individual wall urinals.

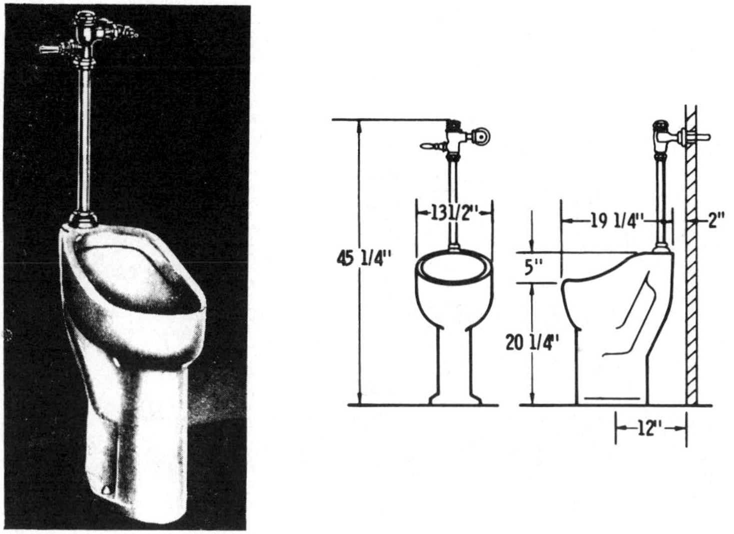

Courtesy Crane Co.

Fig. 8-48. Installation dimensions for a vitreous china pedestal-type urinal.

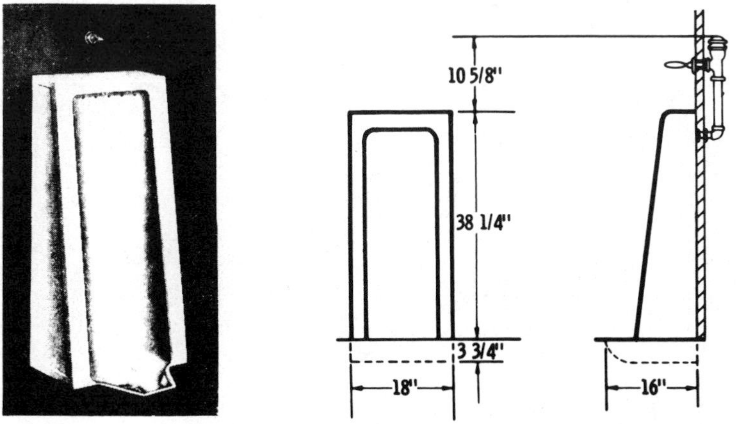

Courtesy Crane Co.

Fig. 8-49. A sloping front, back inlet stall urinal with an integral flushing rim. Note the concealed flush valve with oscillating handle.

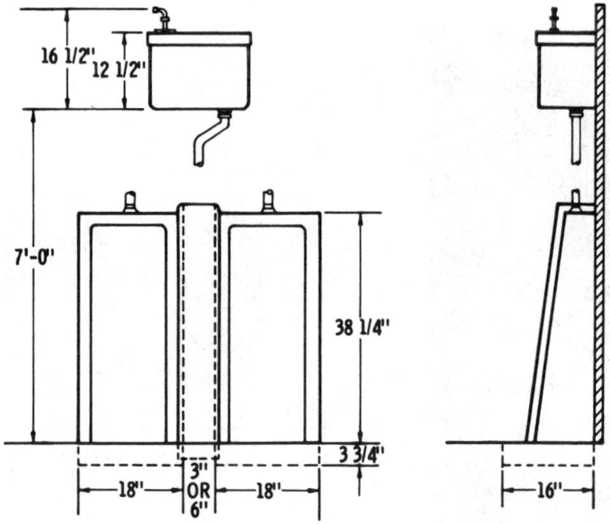

Courtesy Crane Co.

Fig. 8-50. Installation dimensions for a battery of two stall-type urinals.

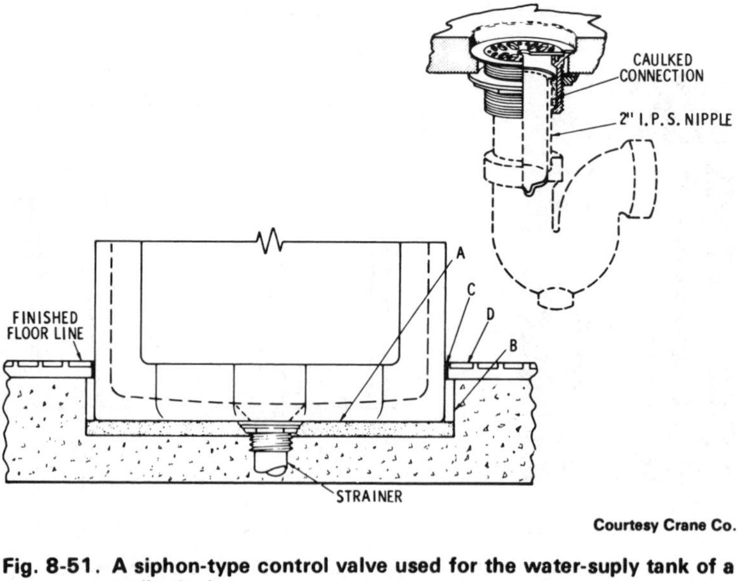

Courtesy Crane Co.

Fig. 8-51. A siphon-type control valve used for the water-suply tank of a stall urinal.

the finished floor and the stall (see Fig. 8-51C), and filled with a plastic waterproof compound.
4. The finished floor should be sloped to drain into the lip of the urinal (see Fig. 8-51D).

A siphon-type control valve (Fig. 8-52) is used in the supply tank for some urinals. The water is removed from the supply tank by siphoning action, and there are no moving part. A flushing action occurs at intervals of one to sixty minutes, depending on the rate of tank refill. An eccentric outlet permits use of the valve in tanks having either an end outlet or a center outlet.

A float-type valve (Fig. 8-53) is also used in the water supply tank of stall urinals. The water supply can be throttled without clogging. Since the valve is never completely shut off, a stop should be provided in the supply line. A single screw with lock nut controls the discharge rate. The valve can be adjusted to fill the tank at at intervals of one to sixty minutes.

As mentioned previously, tne latest manufacturer's instructions should be followed for a given installation. Dimensions for typical flush pipe assemblies for urinals with high water supply tanks are given in Fig. 8-54.

BIDET

The *bidet* (pronounced be-day') is a relatively new fixture in this country; however, it has been in common use in the Latin American countries and in Europe for many years (Fig. 8-55). This fixture can be a valuable contribution to personal cleanliness and sensible living for every member of the family. The bidet is designed for cleanliness of the localized parts of the body, and it serves many useful purposes. Its use is a clean habit for men, women, and children. Its frequent application is advisable for comfort and health and in keeping with a mode of sanitary living.

Americans who have traveled either in Europe or in South America have come to accept the bidet as a standard bathroom fixture, and they have learned that it is a logical twin fixture to the toilet and a remarkable aid to personal cleanliness. Since more Americans travel abroad and become acquainted with the advan-

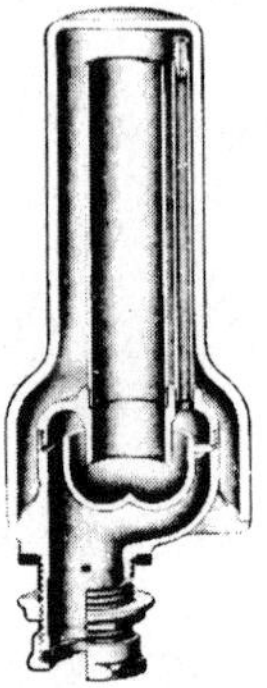

Courtesy Crane Co.

Fig. 8-52. A siphon-type control valve used for the water-supply tank of a stall urinal.

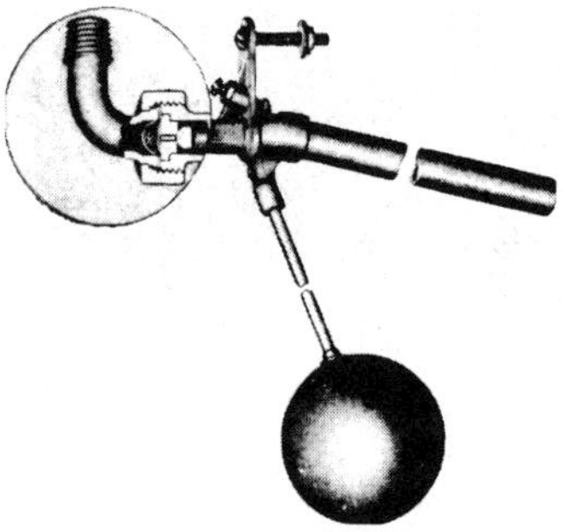

Courtesy Crane Co.

Fig. 8-53. A float valve used to set the rate of water-supply-tank discharge. The valve can be adjusted to fill the tank at intervals of one to sixty minutes.

tages of the bidet, more of these fixtures are being installed in new homes. Doctors who emphasize the basic importance of genitourinary cleanliness are often recommending the bidet and the washing practices made possible by the bidet to their patients of both sexes. In the near future, the bidet may be accepted as a necessary bathroom fixture.

The bidet is equipped with valves for both hot and cold water and with a pop-up type waste plug either for retaining the water or for draining it as desired. The inside walls of the bowl are washed by a flushing rim that uses the same basic principle of operation as the toilet bowl; however, the bidet is neither designed nor intended to carry away human waste material.

The fixture is also provided with an integral douche or jet, operated when desired by means of a transfer valve which directs

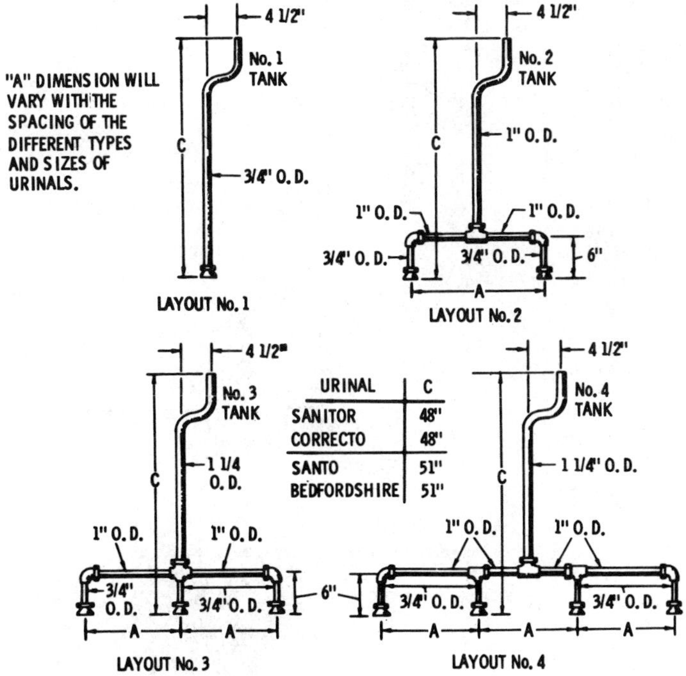

Fig. 8-54. Dimensions for typical flush pipe assemblies for urinals with high water-supply tanks.

a stream or column of water upward from the bottom section of the bowl (Fig. 8-55). This jet is formed by a cluster of small holes which are arranged to direct the water to a central point, thereby forming a solid stream of water.

The bidet is also used as a foot bath. Since the distance from the floor to the top of the rim is 14 to 15 in., it can be used very comfortably as a foot bath. Since the bidet is made from white

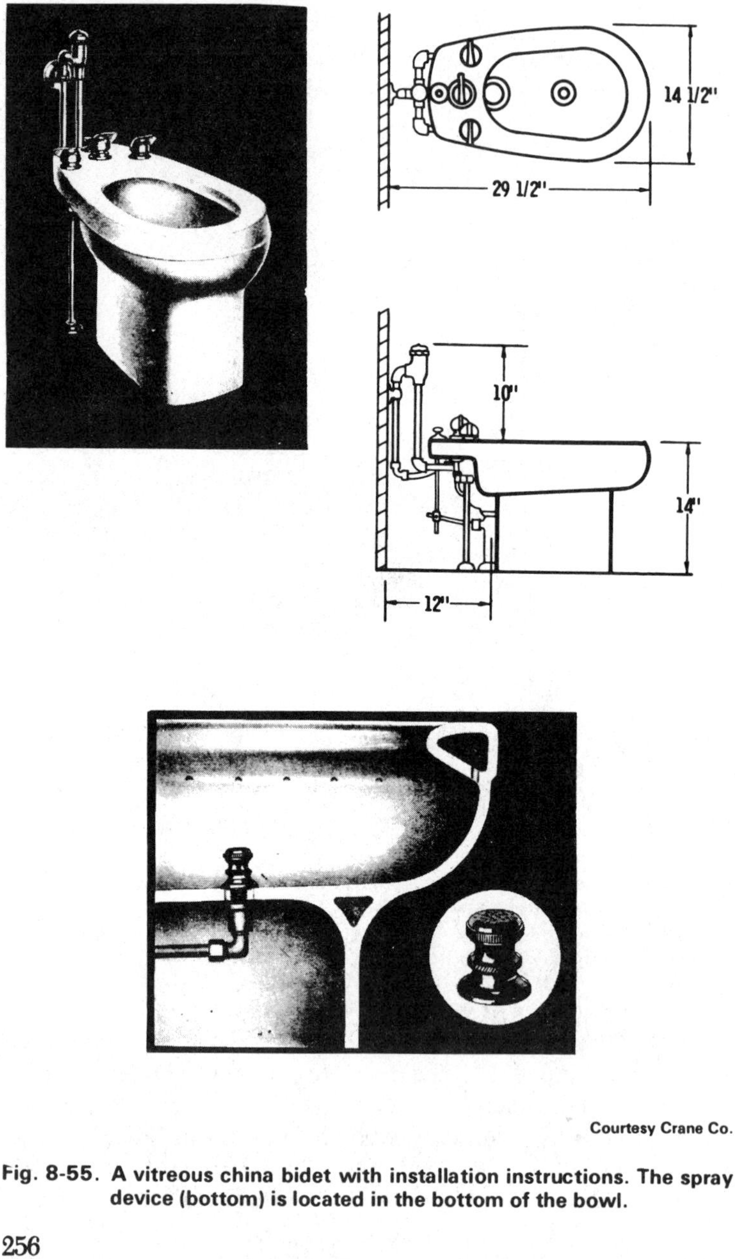

Courtesy Crane Co.

Fig. 8-55. A vitreous china bidet with installation instructions. The spray device (bottom) is located in the bottom of the bowl.

vitreous china, there should be no hesitancy in using it as a foot bath; by means of the flushing rim, it can be kept clean and sanitary by merely rinsing with water or wiping with a damp cloth.

Physicians may advise the use of the bidet for individuals who cannot use the bathtub or shower because of ill health; it helps elderly people to bathe without exertion. The thermal effect and soothing action created by water under pressure is often recommended for irritations of the skin or following operations or injuries to the pelvic area.

CHAPTER 9

Valves and Faucets

Valves are used in a piping system to control the flow of liquids, gas, or air. There are many different types of valves; some of the commonly used types and the reason for their use are explained in this chapter.

Faucets are also valves, but they are valves located at the end of a main or branch and serve to control the flow of liquids at a terminal point or at a fixture.

VALVES

Gate valves are most commonly used in industrial and commercail applications as stop valves, to turn on or to stop flow as opposed to regulating flow. They are usually used to control a main, branch, or an item of equipment. A gate valve has a full-size passageway when fully open and is normally used in either a full open or full closed position. Gate valves are best suited for full open flow because the fluid moves through them in a straight line, nearly without resistance or pressure drop, when the disc is raised from the waterway. Gate valves are not intended for throttling the flow or for frequent operation. If the valve is only partially open, it may vibrate or chatter and cause damage to the seating surface. Repeated movement of the disc near the point of closure will cause the seating surfaces to rub together until they are galled or scored on the downstream side. There are two basic types of gate valves: *rising stem* and *nonrising stem.*

Rising stem valves are generally installed where adequate space is available and in conditions where the visual determination of whether the valve is open or closed is important.

Nonrising stem valves are generally installed where space is limited. Because the disc rides up and down on the stem and the stem rotates in the packing, the life of the packing is greatly increased. However, it is not possible to determine visually whether a nonrising stem valve is open or closed.

Gate valves are made with solid wedge discs (Fig. 9-1) and double wedge discs (Fig. 9-2). They are also made in rising stem and nonrising stem patterns. A gate valve is normally used in either a full open or fully closed position. The flanged valve (Fig. 9-1) is an O.S. & Y. (open screw and yoke) type.

Globe Valves

About the most familiar type of valve is the globe valve, which is extensively used in piping systems for water, air, and steam. This type of valve is designed to be placed in the run of a pipe line. As shown in the cutaway view in Fig. 9-3, a globe valve has a spherical casting with an interior partition that shuts off the inlet from the outlet except through a circular opening in the seat. Screwed into an opening in the top of the casting is a plug having a stuffing box and a threaded sleeve in which the valve spindle operates. On the lower end of this spindle is the valve proper, and on the other end is a handwheel. The valve is closed by turning the handwheel clockwise, which lowers the spindle and valve until it presses firmly and evenly on the valve seat, thus closing communication between the inlet and outlet. By turning the handwheel in the opposite direction (counterclockwise), the valve is opened.

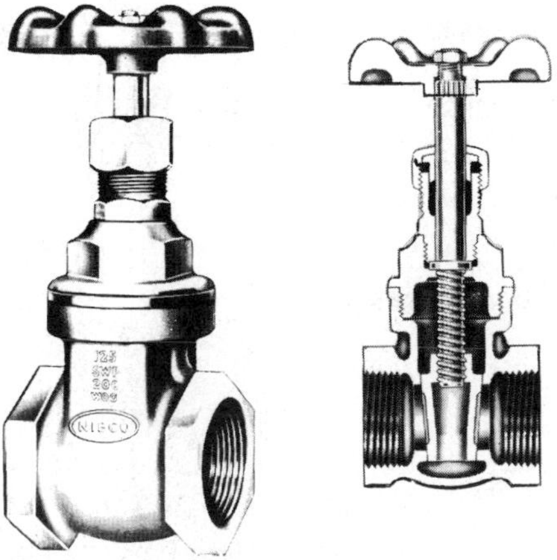

GATE VALVE
NON-RISING STEM—
SOLID WEDGE DISC

(A) I.P.S. (threaded).

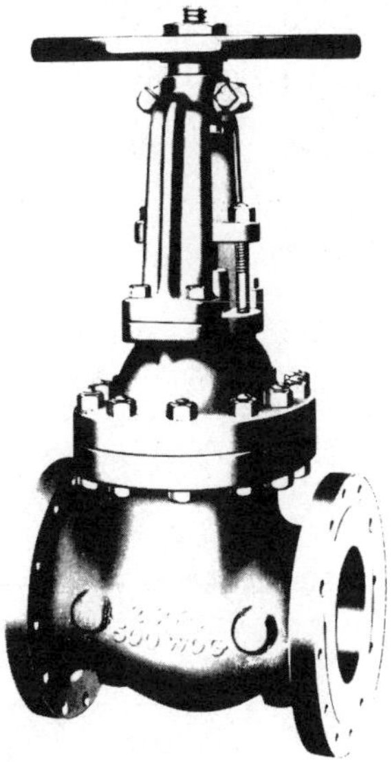

(C) Flanged.

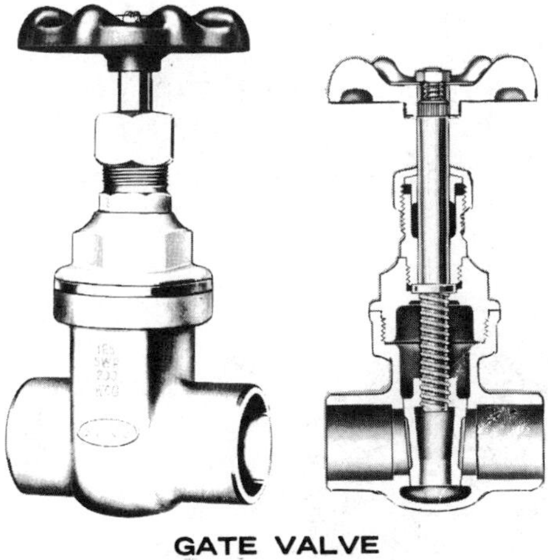

GATE VALVE
NON-RISING STEM—
SOLID WEDGE DISC

(B) Copper to copper (sweat).

Fig. 9-1. Bate valves: (A) I.P.S. (threaded); (B) copper to copper (sweat); (C) flanged.

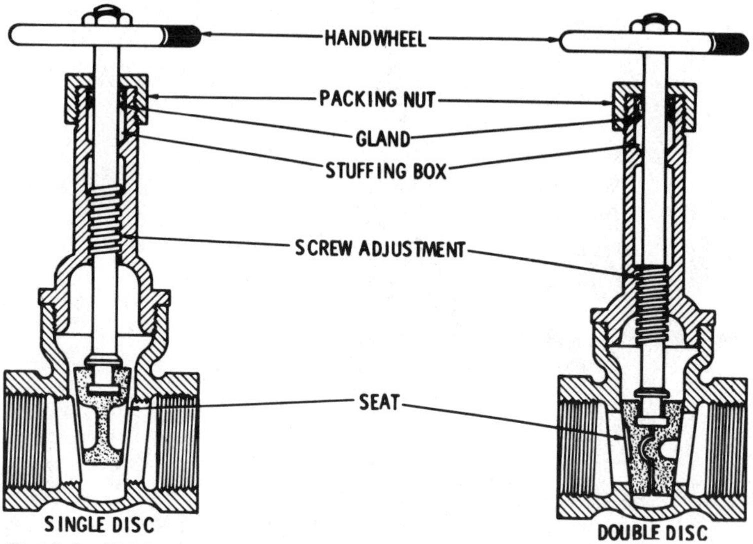

Fig. 9-2. Single- and double-disc gate valves.

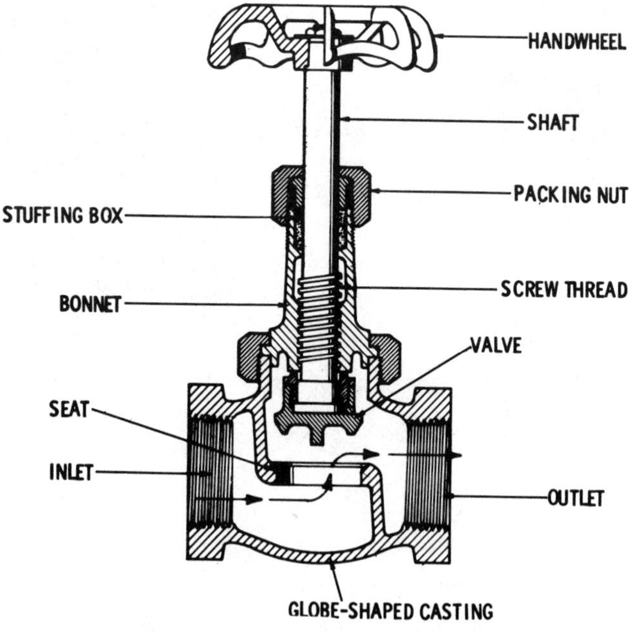

Fig. 9-3. A globe valve.

Globe valves are primarily control valves because of their throttling characteristics. The disc travel is shorter, and they generally operate with fewer turns of the handwheel than that of a gate valve. The maintenance of these valves is most economical since they do not need to be removed from the pipe line for disc replacement. Steam discs and W.O.G. (water-oil-gas) discs are securely fastened to a swivel-type holder to ensure even wear and uniform seating. Globe valves should be installed with the pressure on top of the disc, whenever conditions permit, so that the line pressure can add to the seating pressure. This is particularly useful in hot-water and steam applications where subsequent cooling of the valve may cause leakage to occur due to the different cooling rates of the various parts of the valve. Service conditions often make installation with the pressure under the seat desirable; and in some cases, such as boiler-feed water lines, such installation is mandatory by codes or regulations.

Fig 9-4 shows two major types of globe valves: the I.P.S. (threaded) and the copper to copper (sweat).

The seat and valve may have their contact surfaces either flat or beveled, as shown in Fig. 9-5. The valve disk may be of metal or fiber. Fiber seats should be interchangeable. A globe valve will remain leakproof longer than a gate valve. An objection is that unless properly designed, the opening through the seat of the valve is not the full area of the pipe size; this and the contorted passages offer considerable reisitance to water flow. A serious objection on water lines that must be drained in freezing weather is that it is impossible to drain the water from a horizontal line when the valve spindle is in an upright position. In piping up such lines, always have the spindle horizontal, as shown in Fig. 9-6.

Angle Valves

An angle valve is virtually a *glove valve with the inlet and outlet at* 90° to each other, as shown in Fig. 9-7. Such valves can serve the double purpose of controlling the flow and changing the direction of the pipe line, thus eliminating the need of an elbow. Angle valves are also make with either metal seats or soft seats; the latter should be used on water lines.

In connecting a globe-type valve, it is important to place it in the

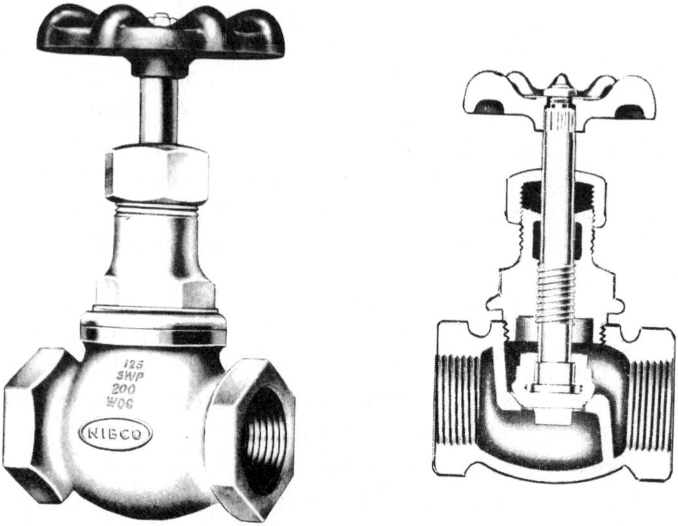

GLOBE VALVE
I. P. S. (THREADED)

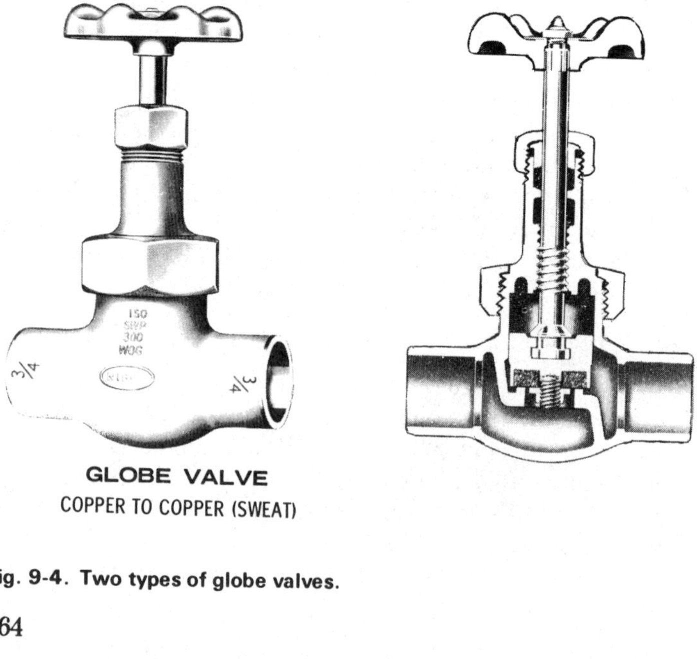

GLOBE VALVE
COPPER TO COPPER (SWEAT)

Fig. 9-4. Two types of globe valves.

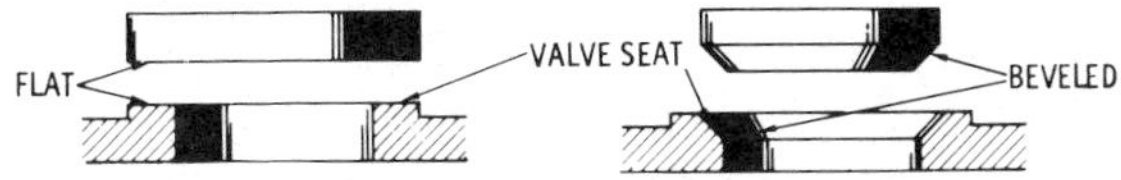

Fig. 9-5. A flat and a beveled valve and valve seat.

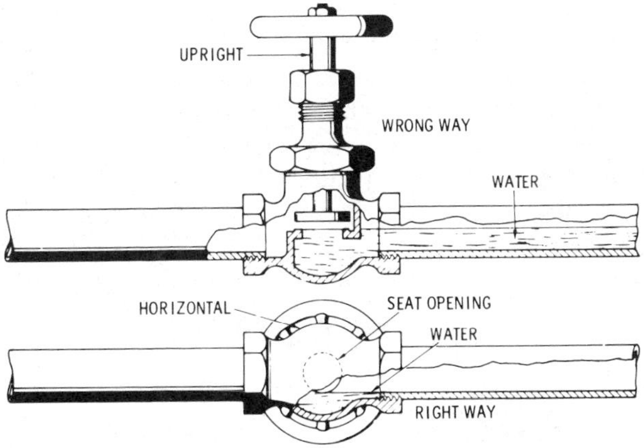

Fig. 9-6. The proper way to install a globe valve.

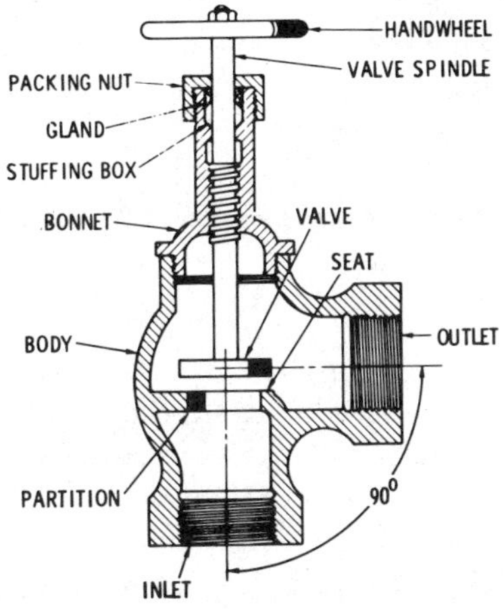

Fig. 9-7. An angle valve.

line so that its inlet side will carry the pressure when the valve is closed, otherwise it will be impossible to repack the stuffing box while the line is under pressure.

Ball Valve

One of the newer valves on the market is the ball valve (Fig. 9-8), so called because it has a hollow ball inside. When you manipulate the lever handle, the ball covers or uncovers holes to let the water flow freely or to stop it. Of all the common valves, the bell valve has the most positive shutoff and the interior is the smoothest—it's coated with Teflon. It does not wear out as quickly as a gate or globe valve (though they last years). In the home, valves on runs or pipe are constructed either of cast brass or plastic.

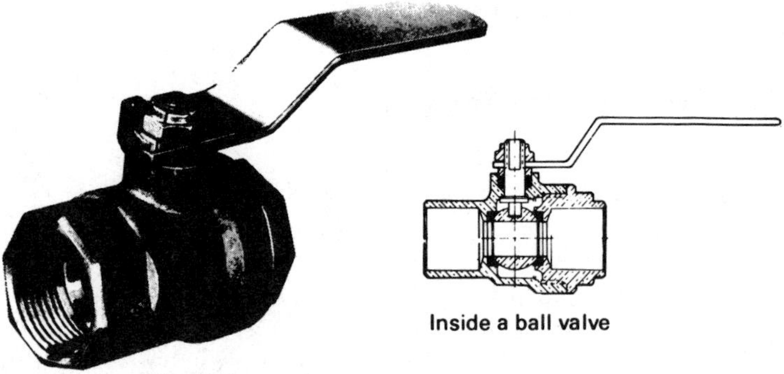

Inside a ball valve

Ball valve

Fig. 9-8. A relative newcomer on the scene is the ball valve.

Cross Valves

The essentials of a cross globe valve are shown in Fig. 9-9. This type of valve is used where it is desired to control the flow at the junction of a main line and a branch. In cases where the branch is the inlet, the valve can be repacked while under pressure, but unfortunately the branch must frequently be made the outlet, in which case the valve cannot be repacked while under pressure. The operation of regrinding is performed in the same way as for a globe valve.

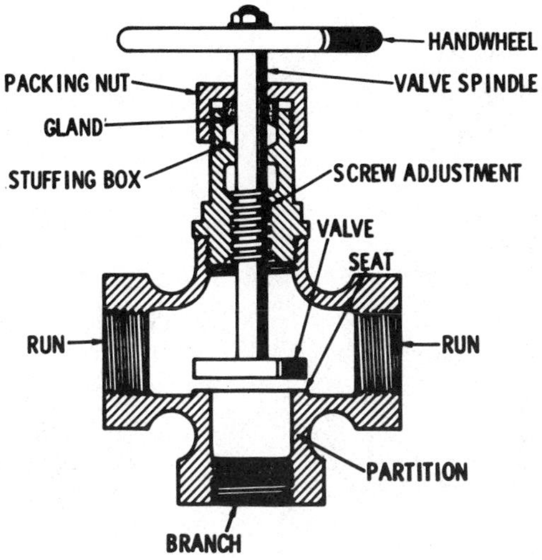

Fig. 9-9. A cross globe valve.

Compression Stops

Compression stops are essentially globe valves and are commonly used on the supply piping to individual fixtures. Most plumbing codes require valves on the supply piping to individual fixtures so that when the faucet needs repair it will not be necessary to shut off the water to the entire building in order to make the repair. Compression stops are made in straight angle patterns, and for I.P.S. and sweat connections. They are also made in rough brass, polished brass, and chrome-plated finished, depending on the usage of the stop. Several types of compression stops—the kind usually used to control sinks and toilets—can be seen in Fig.9-10.

Stop and Waste Valves

Stop and waste valves are used primarily to protect piping from freezing. A stop and waste valve is essentially a compression stop with a drain feature built into it. When the valve is closed and the button and rubber washer are removed, by turning the button counterclockwise, any water in the piping from the valve to the end of the piping will drain out, if the piping is open on the end.

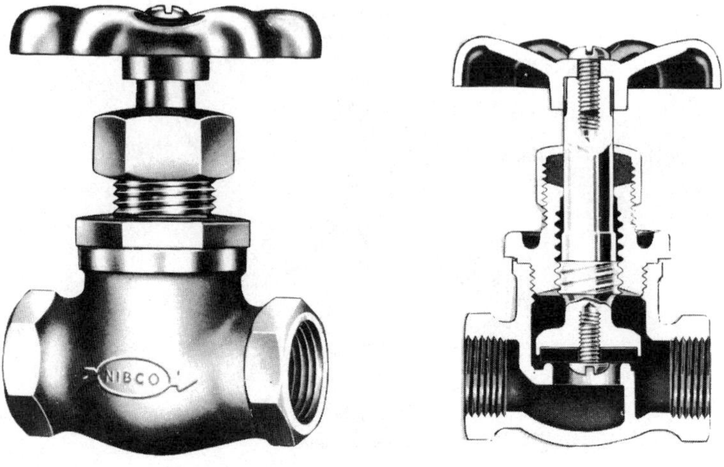

STOP VALVE
I.P.S. (THREADED)

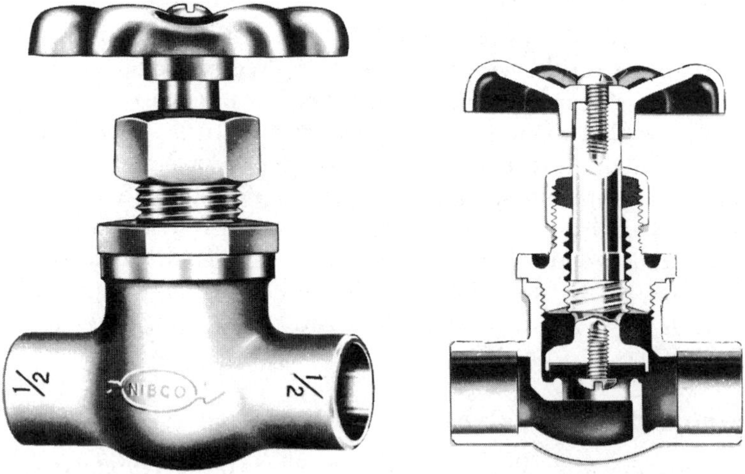

STOP VALVE
COPPER TO COPPER (SWEAT)

Fig. 9-10. Four different

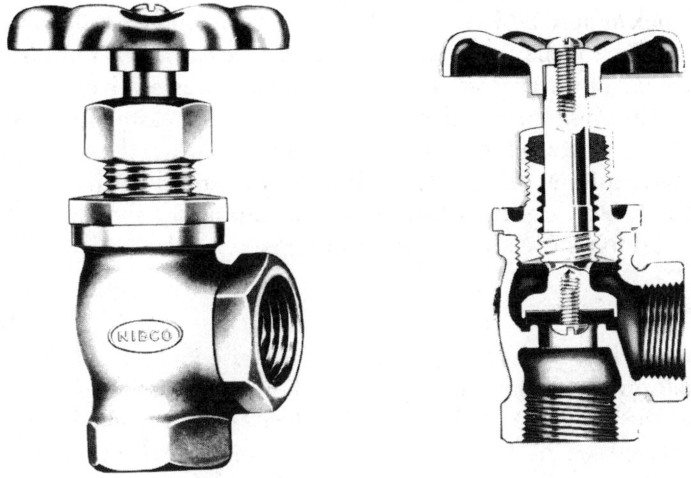

ANGLE STOP VALVE

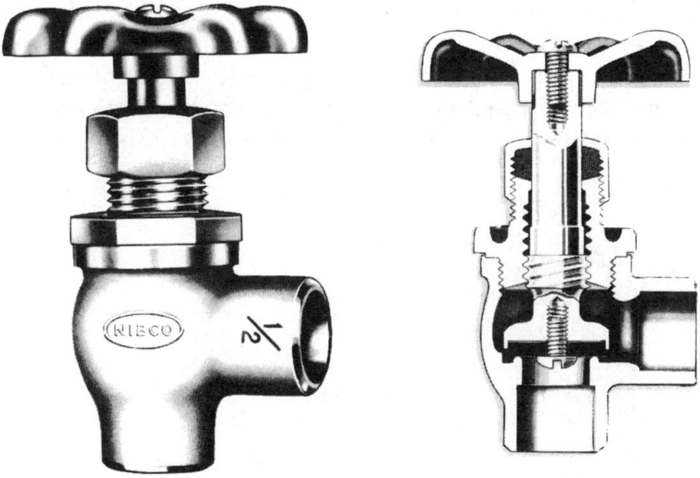

ANGLE STOP VALVE
COPPER TO COPPER (SWEAT)

compression stop valves.

The common usage for a stop and waste valve is to protect a sillcock or hose bibb. It is usually necessary to open the sillcock or hose bibb to relieve any vacuum that may be caused by the partial drainage of water from the piping. Stop and waste valves are also made in a pattern called an automatic stop and waste. Automatic stop and waste valves have a port that opens when the valve is in closed position, thus the valve drains automatically. A hose bibb, sillcock, etc., which is on the end of the piping run should also be opened to relieve a possible vacuum.

Needle Valves

A needle valve is a form of globe valve used where only a very small amount of flow and close regulation are desired. In place of a disc, the pointed end of the spindle forms the valve, which sets on a beveled seat of the same taper. The standard angle of the seat is 30° to the spindle axis. Fig. 9-11 shows the construction of a needle valve.

Pressure-Regulating Valves

Pressure-regulating valves (Fig. 9-12) are used for many purposes by the plumbing and pipe fitting trades. The principal uses are on water, air, and oil service piping. The basic principles of valve operation are the same regardless of the type of usage.

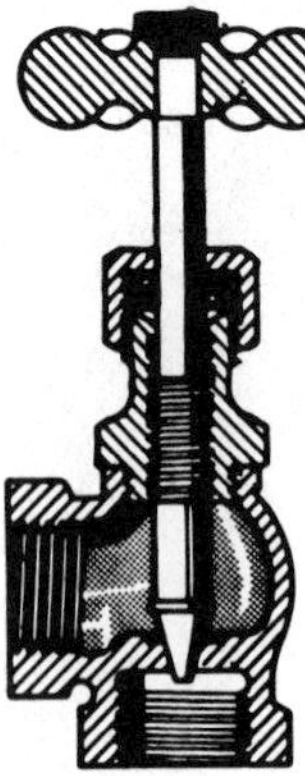

Fig. 9-11. A needle valve.

Fig. 9-12. A high-capacity water-reducing valve with integral strainer.

Courtesy Watts Regulator Co.

When pressure-regulating valves are used on water piping, they may be used to maintain a lower than supply main pressure in a building. If, for instance, the supply main pressure would be 125 psi it would be desirable to lower the building pressure to 60 psi to protect the piping and appliances. It is recommended by many manufacturers of commercial type dishwashing equipment that the working pressure of the dishwasher not be over 25 psi. A pressure-reducing valve can be installed between the booster heater and the dishwasher for this purpose. A pressure-reducing valve should have an integral strainer (Fig. 9-13), or a strainer should be installed on the high-pressure or inlet side of the valve.

Float Valves

The duty of a float valve in to shut off the water supply to a tank or receptacle when the water has reached a predetermined level. The automatic action is due to the rising level of the water during the filling of the tank, carrying up with it a float which, by suitable gearing, closes the water supply valve. When the water is discharged from the tank in flushing, the float descends by gravity, and the valve opens by pressure of the water supply.

The essentials and operation of a float valve are shown in Fig. 9-14. In practice, there is a great variety of float valves, all acting on the basic principles shown but having various modifications of the transmission gearing. Usually, some means of adjusting the

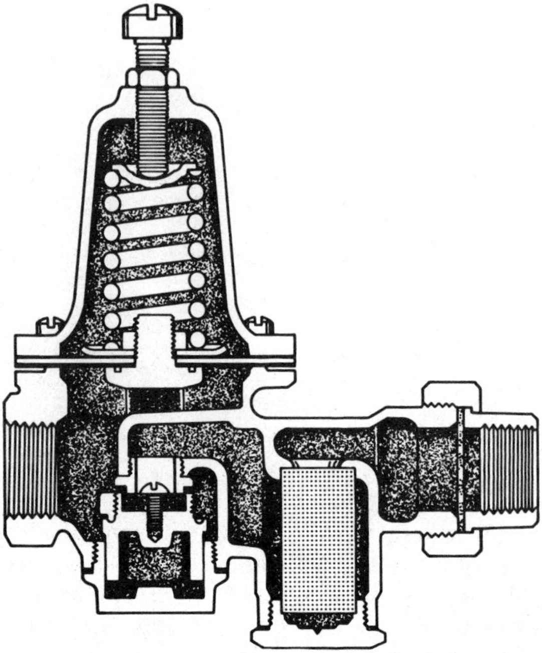

Fig. 9-13. Cutaway view of a standard-capacity water-reducing valve with integral strainer.

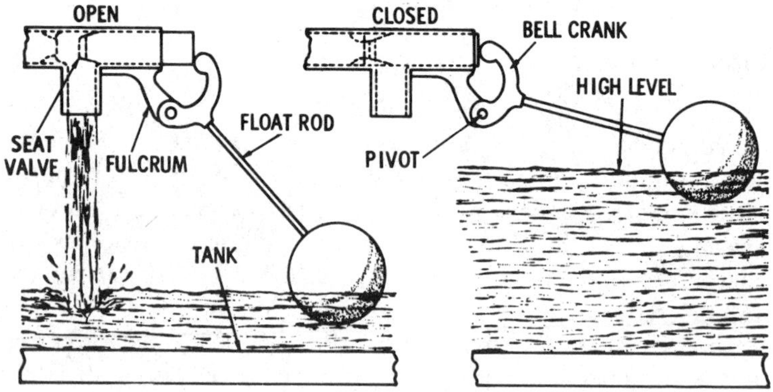

Fig. 9-14. Float valve in open and closed position.

water level at which the valve closes is provided. The principle of such adjustment is shown in Fig. 9-15, and use should be made of this means of adjustment when necessary rather than resorting to the objectionable practice of bending the float rod. Fig 9-16 shows how the rod is bent to lower the level from *A* to *B*, and the possible result due to such practice.

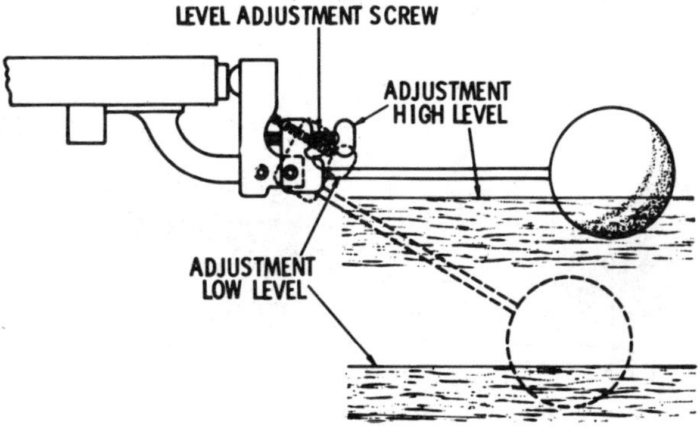

Fig. 9-15. A float valve level adjustments.

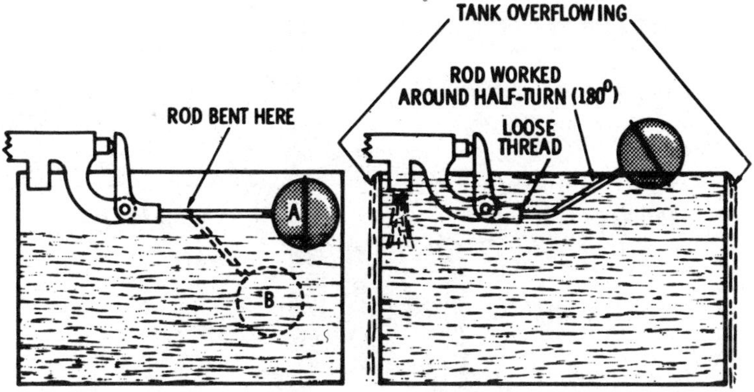

Fig. 9-16. A method of adjusting the float valve by bending the rod.

A form of valve employing a diaphragm is shown in the open and closed position in Fig. 9-17. It acts by hydraulic pressure, depending on the differential area principle to keep it closed. In construction, a diaphragm valve divides the valve chamber into two compartments. In the lower compartment is the valve seat of the diaphragm valve, outlet ports, and a bypass passage leading to the upper compartment that has no opening except that of an auxiliary valve operated by the float as shown. This arrangement forms an efficient device and can be made to control the flow from a large tank with a small float. Fig. 9-18 shows the shank of the float valve and the method of securing a tight joint where the shank passes through the bottom of the tank.

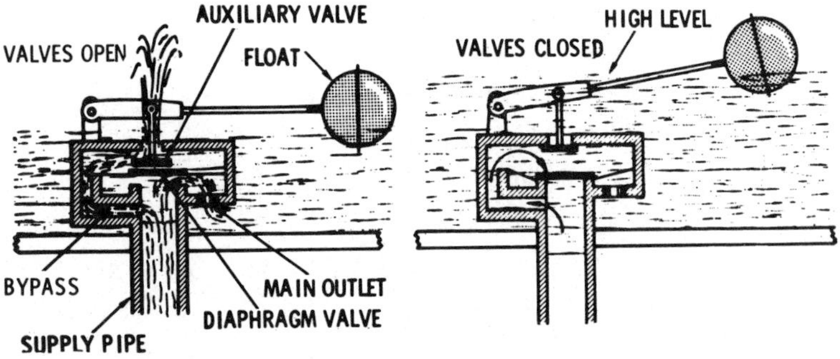

Fig. 9-17. Adjustment of a hydraulic float valve.

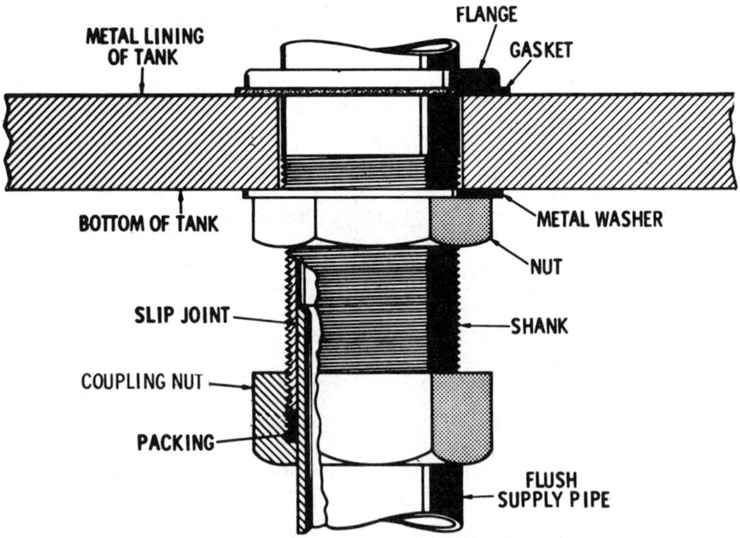

Fig. 9-18. The base of a fill valve.

Flush Valves

Flush valves are used primarily on water closets and urinals. One type is shown in Fig. 9-19. This diaphragm type requires no regulation to maintain flushing accuracy. When the valve is in closed position, the segment diaphragm divides the valve into an upper and lower chamber with equal water pressures on both sides of the diaphragm. The greater pressure area on top of the diaphragm holds it closed on its seat. Movement of the handle in any direction pushes the plunger, which tilts the relief valve and allows water to escape from the upper chamber. The water in the lower chamber, below the diaphragm, now being greater, raises the entire assembly of working parts, the relief valve, disc, diaphragm, and guide as a unit and allows the water in the lower chamber to spill over into the outlet and flush the fixture.

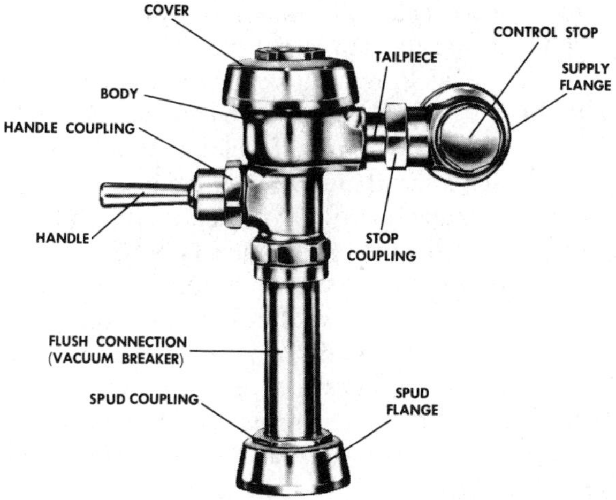

Fig. 9-19. Diaphragm flush valve.

When the flush valve does not shut off, it is usually due to the bypass in the diaphragm being clogged. To check or replace the diaphragm, turn the control valve off, and use a smooth jaw wrench (Fig. 9-20) to turn the chrome-plated nut on top of the cover counterclockwise to loosen and remove the cover. The use

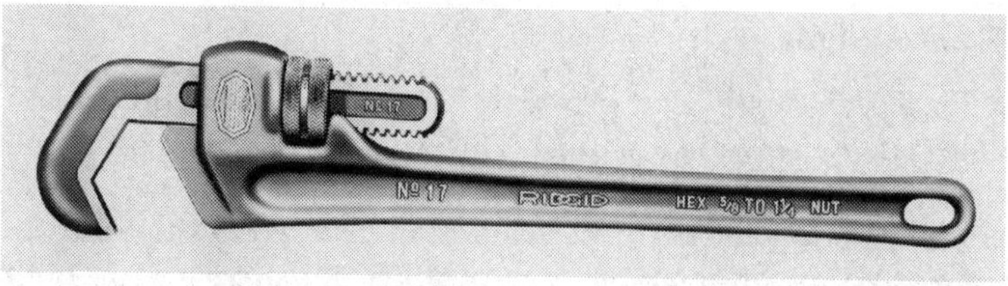

Fig. 9-20. A smooth jaw wrench should be used on chrome-plated fittings to prevent damage to the plating.

of a smooth jaw wrench is advised to prevent damage to the chrome plating. Lift out the inside cover and remove the relief valve, disc, and diaphragm. Inspect the bypass to make certain it is open; if the diaphragm has deteriorated, it should be replaced.

Check Valves

There are many different check valves designed for many different applications. Basically, all check valves are designed with the same purpose: to permit unrestricted forward flow and at the same time prevent backflow. The basic types of check valves are the swing check, the horizontal lift check, and the vertical lift check. Common types of swing check valves are shown in Fib. 9-21 and 9-22. Generally speaking, swing check valves offer less resistance to flow than other types of check valves. The horizontal and vertical lift check valves shown in Fig. 9-23.

Fig. 9-24 shows a swing check valve—the form generally used. The valve should be sufficiently large in diameter to deliver the required amount of water without lifting the disc more than ⅛ in. Higher lifts result in rapid destruction of the valve seat from the hammering action of the valve, especially when used with engine-driven pumps. Of course, with an injector where the feed is continuous, the valve remains off its sear while the injector is in operation, and accordingly, a higher lift is not objectionable.

Temperature and pressure relief valves are specifically designed to protect against excess temperatures and pressures. The valve shown in Fig. 9-25 will relieve if the water temperature rises above 210°F. Standard pressure ratings are 125 and 150 lbs., and pressure build up above the rated pressure of the valve will cause the valve to relieve. The valve shown is design certified by the A.G.A. and is A.S.M.F. rated:

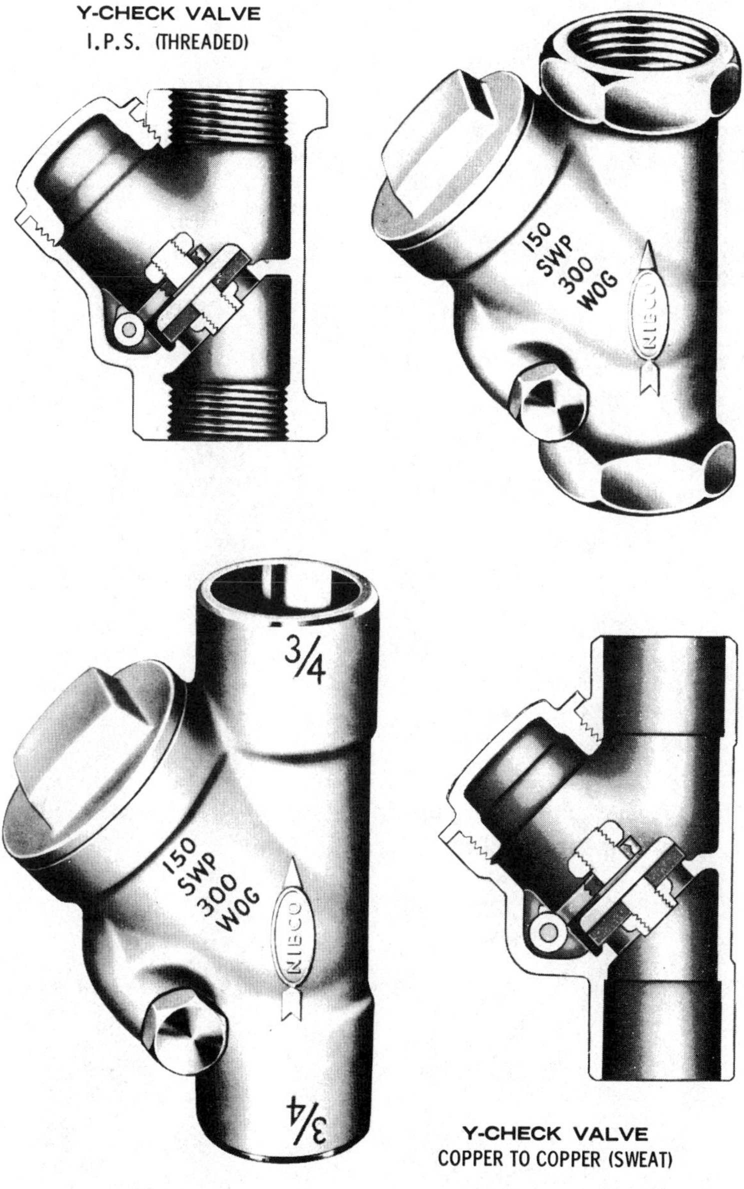

Fig. 9-21. Two common types of swing check valves.

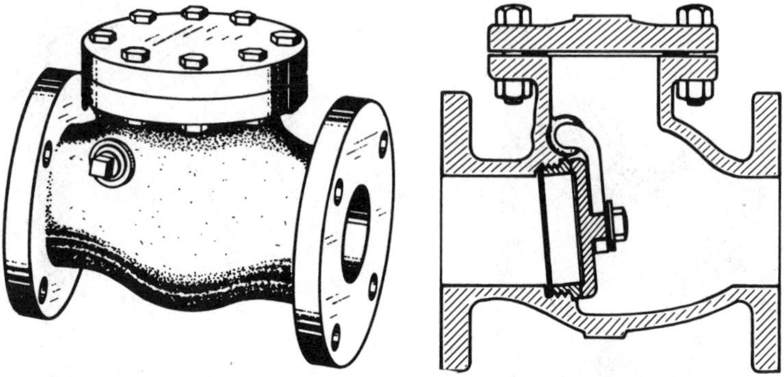

Fig. 9-22. Flanged swing check valve and cutaway view.

(A) Horizontal lift check valve.

(B) Cutaway view of horizontal lift check valve.

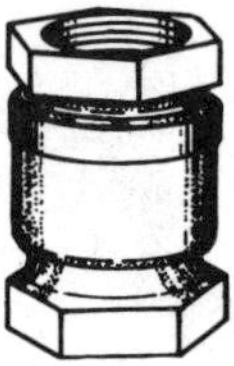

(C) Vertical lift check valve.

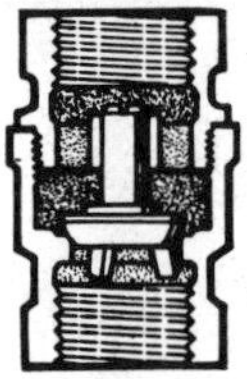

(D) Cutaway view of a vertical lift check valve

Fig. 9-23. Horizontal and vertical lift check valves.

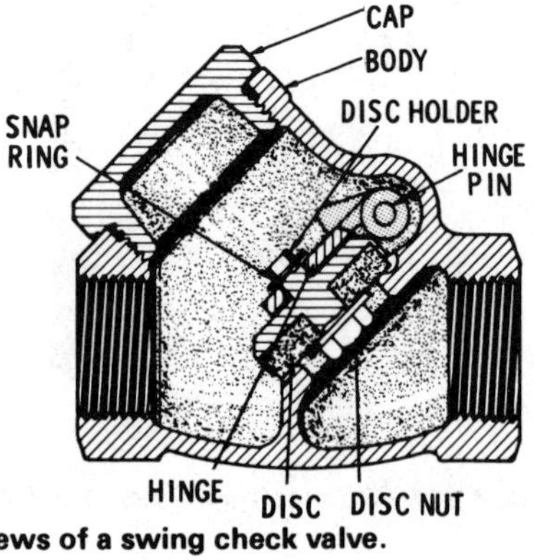

Fig. 9-24. Exterior and sectional views of a swing check valve.

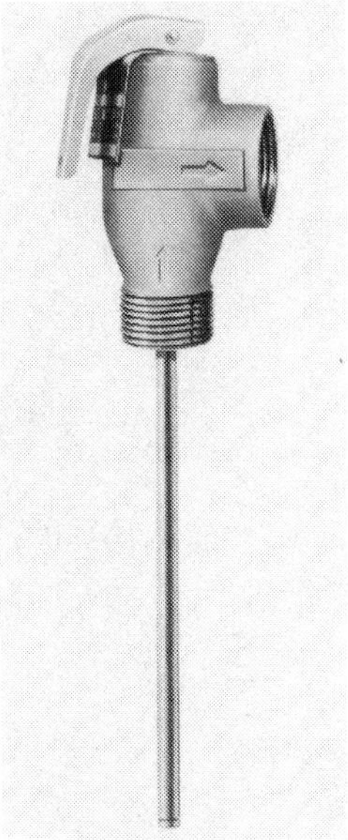

Courtesy Watts Regulator Co.

Fig. 9-25. A temperature and pressure relief check valve.

(A.G.A.—American Gas Association)
(A.S.M.F.—American Society of Mechanical Engineers)

This type of valve is specifically designed for use with hot water storage tanks and heaters.

Safety Relief Valves

A safety relief valve (Fig. 9-26) is the most important device attached to a boiler, as it prevents the steam from rising above the safe working pressure. A good safety valve should be:

1. Large enough in diameter and have sufficient lift to allow the steam to escape as fast as it is generated when the pressure is slightly above that to which the valve is set.
2. It should close as soon as the pressure has dropped a predetermined amount below the set pressure.
3. It should be enclosed or so protected that it cannot be tampered with or accidentally interfered with by contact with foreign objects.
4. For marine purposes, it must be so constructed as not to be affected by the motion of the boat.

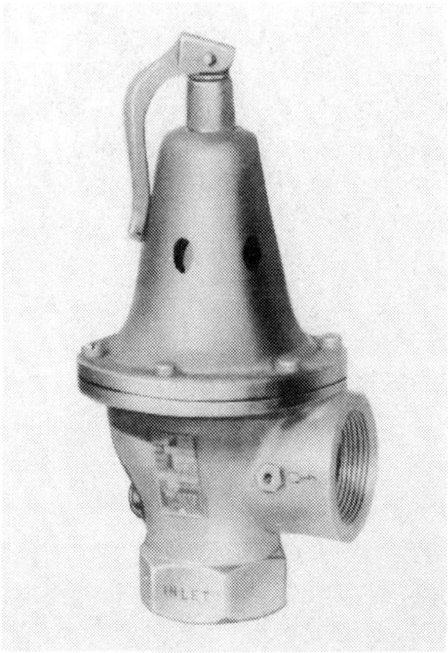

Courtesy Watts Regulator Co.

Fig. 9-26. This A.S.M.E.-rated pressure safety relief valve is specially designed for use on hot-water boilers.

FAUCETS

The terms *faucet* and *bibb* are used generally to signify a valve controlling the outlet of a pipe conveying a liquid. Older faucets use the compression principle of forcing a washer against a metal seat to control or stop the flow of water. More modern faucets use a principle of "shearing off" the flow by partially or completely blocking a port in the body of the faucet. An exploded view of a typical compression faucet is shown in Fig. 9-27. The water flow is controlled by the washer being compressed against the seat by the action of the threaded valve stem when it is turned. Washer damage is the usual cause of a leaky faucet.

Compression Faucets

Compression faucets, although the movement in closing is against the pressure of the liquid, are suitable for moderate and heavy pressures, which is common on city mains. When the valve washer becomes worn or cut, causing the faucet to leak, a new washer can be easily installed. Make sure the replacement is the

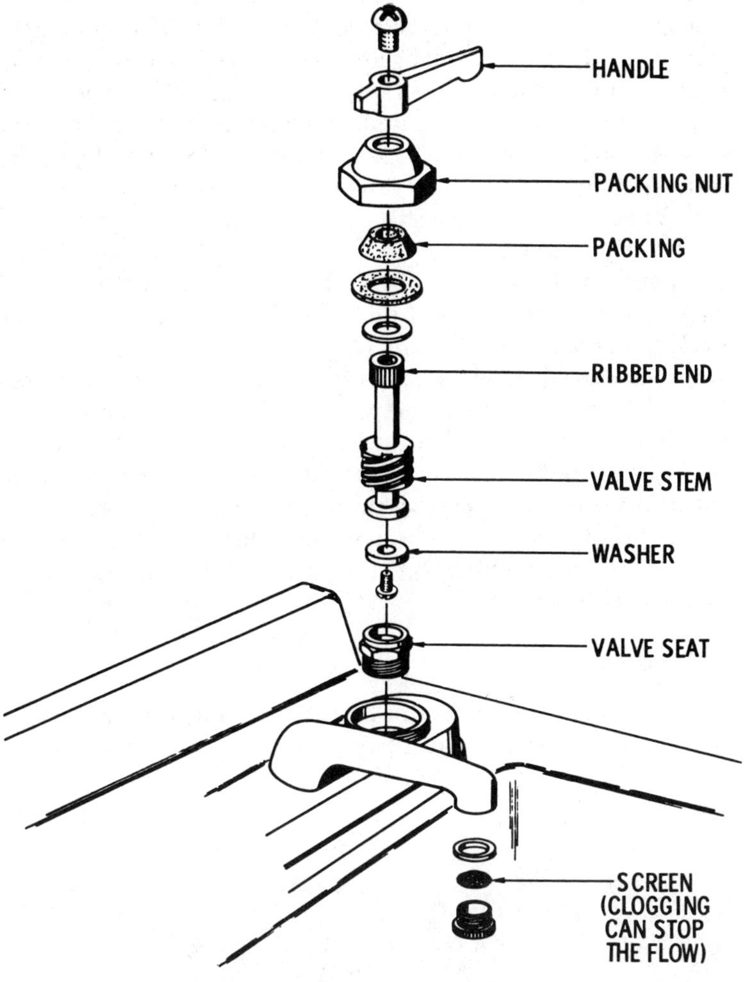

Fig. 9-27. An exploded view of a compression faucet.

same size and type as the old one. Some washers are flat and some are cone shaped, but either type must fit snugly within the retainer on the stem end or it will not work properly. If this retainer is broken, a new valve stem must be substituted or a repair type washer retainer can be used on the end of the stem.

Shut off the water to the faucet, and remove the handle. Remove the packing nut and lift the stem out with it by rotating the stem in the direction normally used to turn the faucet on. Remove the washer with a screwdriver, as shown in Fig. 9-28, being careful not to damage the retainer. Install a new washer; if a cone type, the cone should face outward. Inspect the seat (in the faucet body) for dirt, hard particles, and for roughness and scratches. Clean the seat, if necessary, or replace it if chipped or pitted. The seat screws out for replacement.

Tighten the packing nut if it leaks, but not so tightly that the handle is difficult to turn. If this procedure does not correct the packing nut leak, new packing is required.

In localities where water rates are high, a saving can be effected by installing self-closing compression faucet. This ensures the economical use of water because the valve automatically closes as soon as the handle is released. Self-closing faucets, while economical in the use of water, are objectionable in that the tendency is for the fixtures to receive too little water to keep them in a proper sanitary condition. However, they are used to some degree in public facilities.

Mixing Faucets

Most faucets used on lavatories, bathtubs, and kitchen sinks are of the mixing type. Instead of two separate units—one for hot and

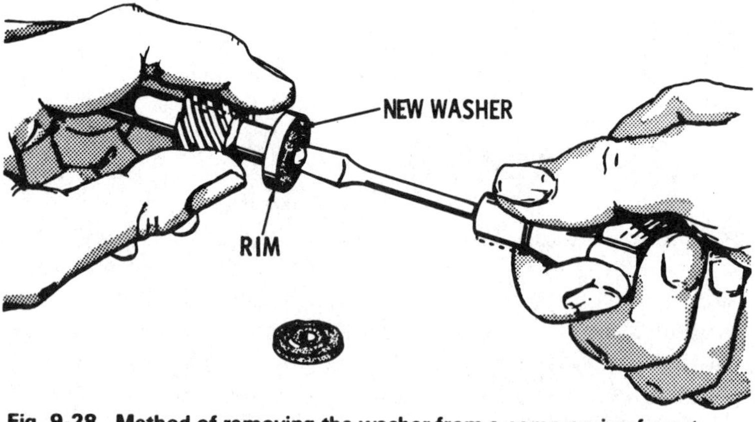

Fig. 9-28. Method of removing the washer from a compression faucet.

one for cold water—the mixing faucet has the hot- and cold-water valves combined with a single spigot. This permits adjusting the temperature of the water to the individual's preference. The installation of this type of faucet is no more difficult than the single types. The repair is similar, the internal construction being very much alike in all cases. An exception to this is the bathtub unit having a shower control valve incorporated, and the kitchen unit having a spray attachment. These additions are relatively simple and seldom need repair or adjustment. Various types of mixing faucets are shown in Fig. 9-29; some of these were illustrated earlier in the book.

Freeze-Proof Faucets

Sill cocks installed through the wall of a house, to make water available for lawn sprinkling and other outdoor activities, are subject to damage during freezing winter weather. Special freeze-proof faucets are available to prevent damage from this cause. As shown in Fig. 9-30, this type of faucet has the valve positioned at some distance from the outlet of the unit. When properly installed, the valve assembly is located well within the warm protection of the house proper—only the outlet and handwheel are exposed to the freezing temperature. When the unit is shut off, the water drains from the extension tube through the spigot, thus preventing freeze-up. Don't be misled by the name, however, for this type of faucet will be as readily damaged by freezing as any other type if the valve-assembly portion of the unit is installed in a location where the temperature will fall below freezing.

Single-Lever Faucets

The single-lever faucet (also shown earlier) combines the convenience of a mixing faucet with only a single control handle. Moving the lever back and to the left turns on the hot water, back and to the right turns on the cold water, while straight back provides a mixture of the two. Any position in between can be selected for the water temperature to suit individual requirements. To shut the water off, the lever is pulled forward in its center position. Repair of this type of faucet is not as complicated as it might seem, usually consisting of the replacement of O-rings

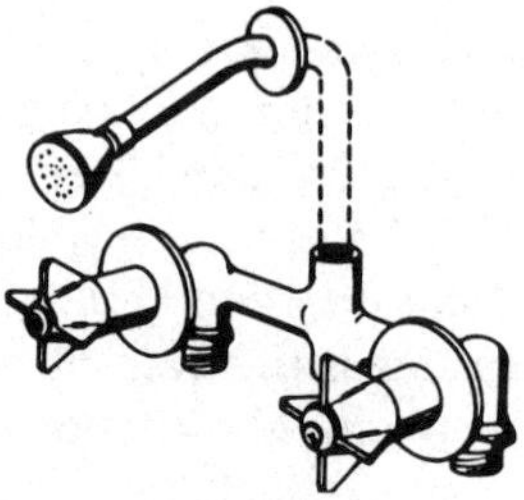

(A) Shower.

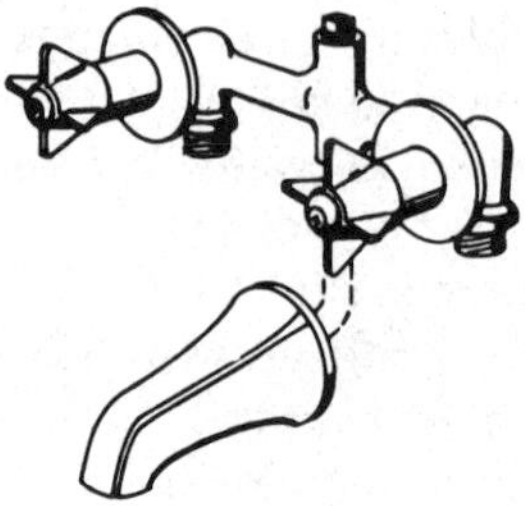

(B) Bathtub.

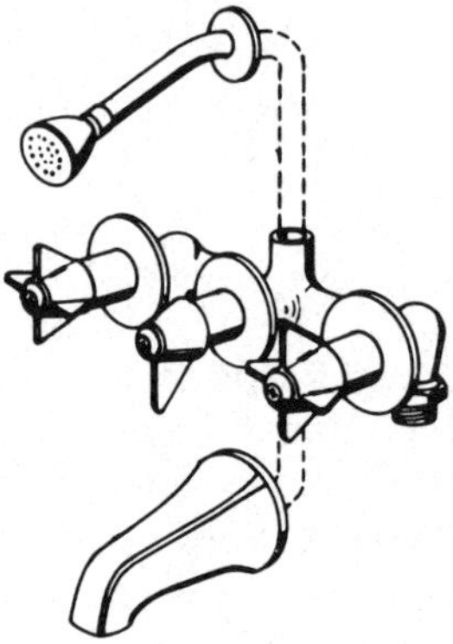

(C) Bathtub and shower.

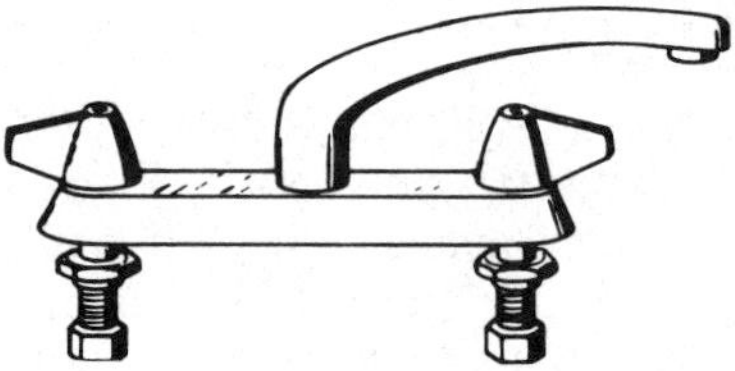

(D) Sink.

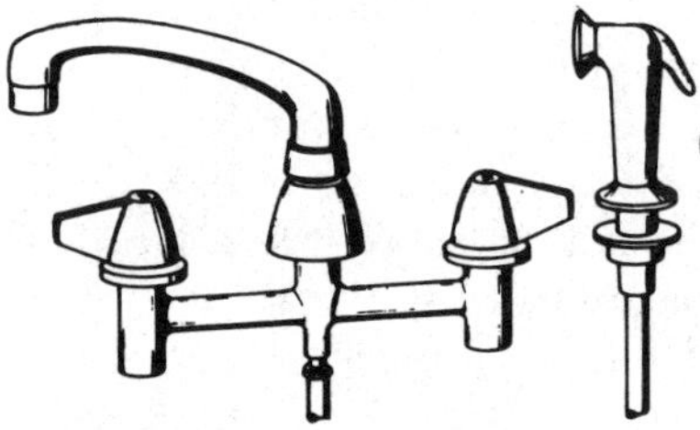

(E) Sink with spray.

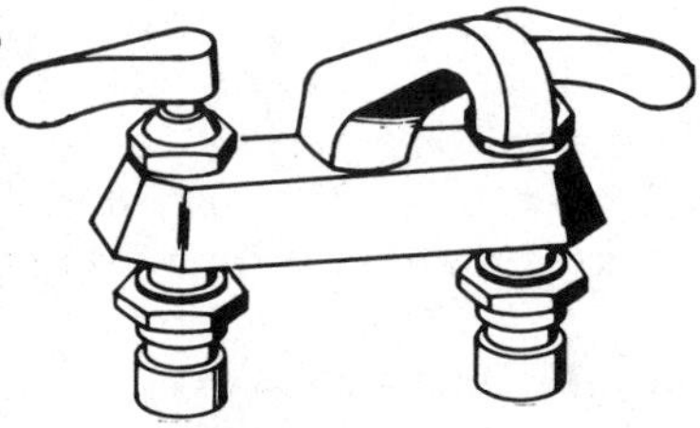

(F) Lavatory.

Fig. 9-29. Mixing faucets.

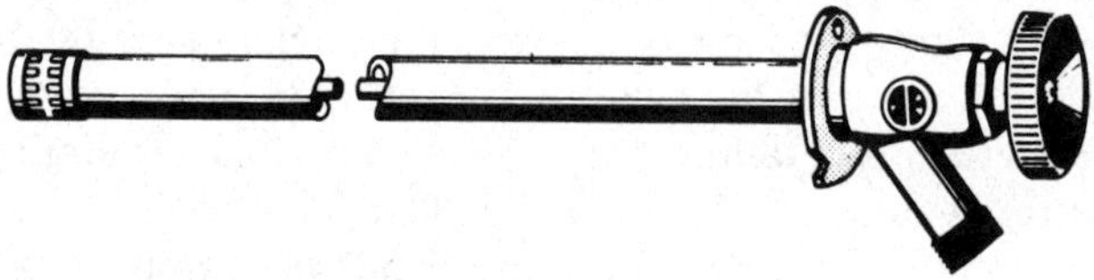

Fig. 9-30. A freeze-proof faucet.

and/or simple rubber valves. But you must use parts only for the brand you have. The exploded view of a typical single-lever faucet in Fig. 9-31 shows the construction details and relative simplicity of this type of unit.

COCKS

A cock is a valve intended to form a convenient means of shutting off the flow of water in a line. It is similar in construction to a ground-key faucet but differs in that it is arranged to be place *in the pipe line* instead of *at an outlet*. The distinction is shown in Fig. 9-32. To meet the various requirements of service, there are several kinds of cocks as follows:

1. Straight-way.
2. Three-way.
    a. Two-port.
    b. Three-port.
3. Four-way.
    a. Two-port.
    b. Three-port.
    c. Four-port.
4. Swing.
5. Waste or drain.
6. Corporation.

Straight-Way Cocks

Fig. 9-33 shows the construction of a straight-way cock, being virtually the same as a ground-key faucet except for the inlet ends and the detachable handle. The general appearance of several straight-way cocks (for steam) is shown in Fig. 9-34. It will be seen that there is a great variety of patterns to meet all requirements.

It should be distinctly understood that the primary duty of a cock is to *control* rather than *regulate* the flow of water, that is, to shut off water from a pipe line in case of repairs or for draining in cold weather. In order to ensure that the cock handle will be turned to the fully open or fully closed position, some units are provided with stops and a check pin. The stops are single projec-

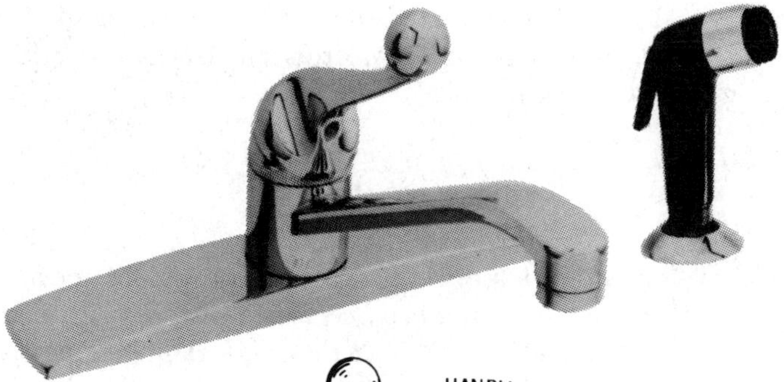

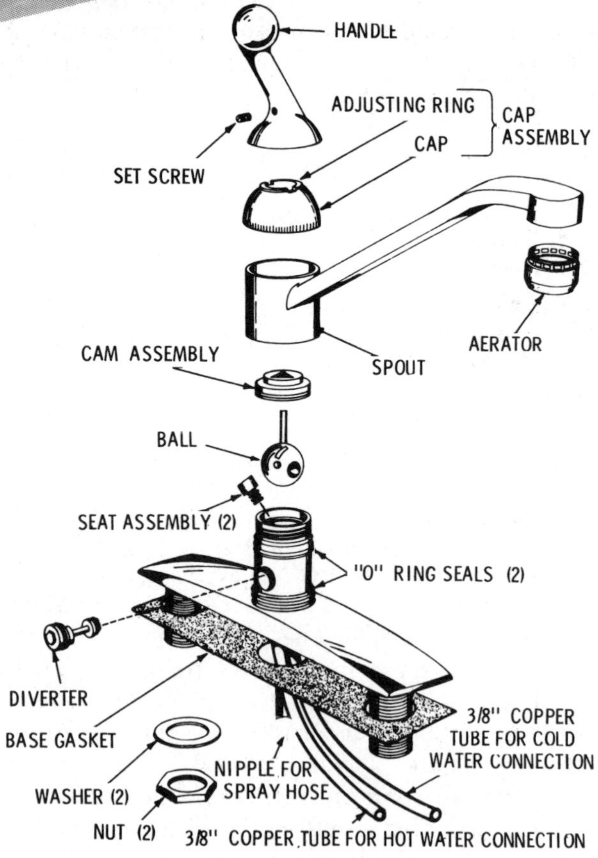

Fig. 9-31. Exploded view of a typical single-lever faucet.

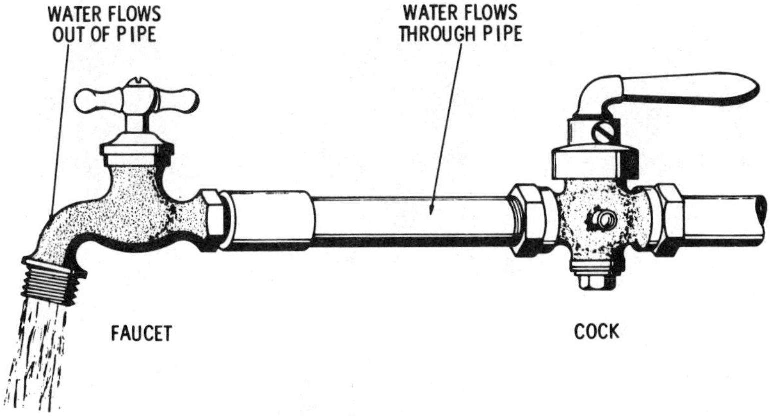

Fig. 9-32. Difference between a faucet and a cock.

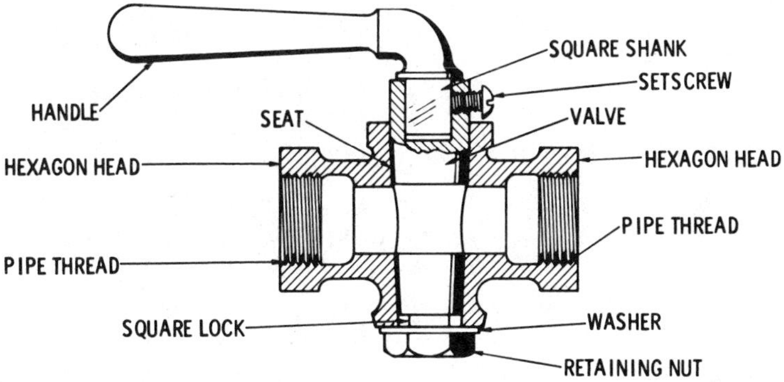

Fig. 9-33. A straight-way cock.

tions on the body adjacent to the valve, and the check pin is inserted in the valve so that it will strike against the stops, limiting the angular movement.

Three-Way Cocks

Three-way cocks are used to control the flow at the junction of:

1. A main line and two branch lines.
2. A main line and one branch line.

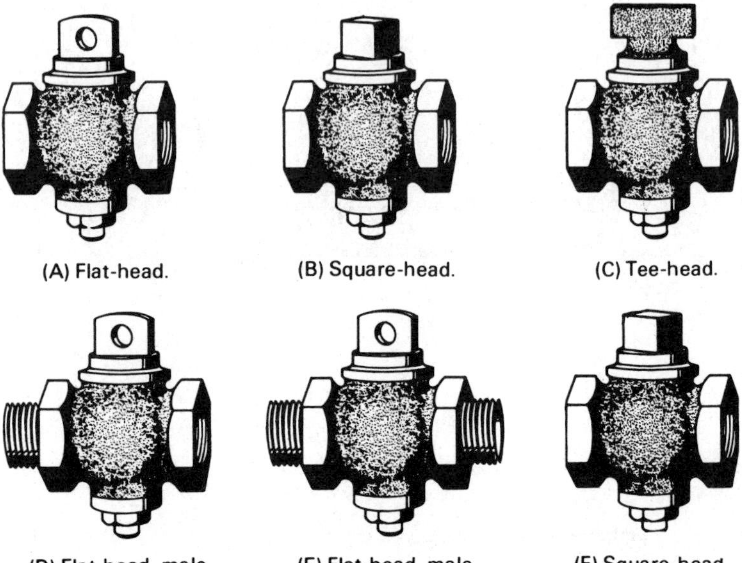

(A) Flat-head. (B) Square-head. (C) Tee-head.

(D) Flat-head, male and female. (E) Flat-head, male both ends. (F) Square-head, with check pins, female both ends.

Fig. 9-34. Various types of straight-way cocks.

For a main line and two branch lines, a two-port three-way cock is used, as shown in Fig. 9-35. The water may be directed to either branch or shut off from both. Where there is only one branch, the three-port cock permits control of the flow to the branch and to the run of the main line beyond the cock, as in Fig. 9-36.

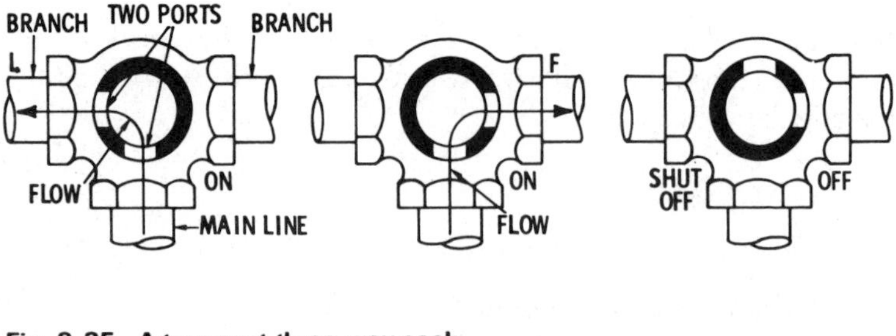

Fig. 9-35. A two-port three-way cock.

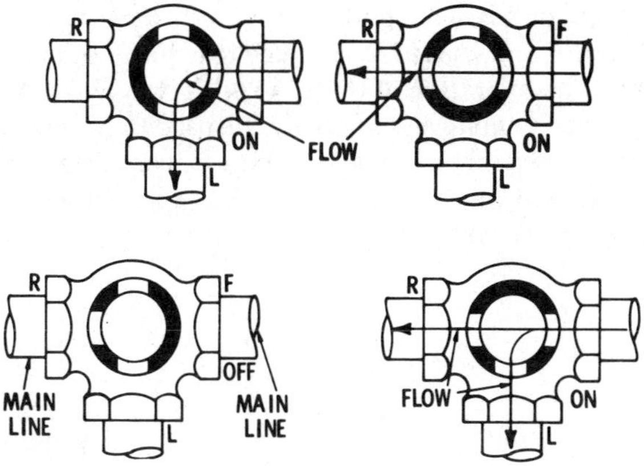

Fig. 9-36. A three-port three-way cock.

Four-Way Cocks

The range of flow control with a four-way cock is quite varied, as this pattern may be had with either two-, three-, or four-port valve is shown in Fig. 9-37.

Waste or Drain Cocks

A waste or drain cock is used to drain a line from which the water is shut off. This is accomplished with a straight-way cock in

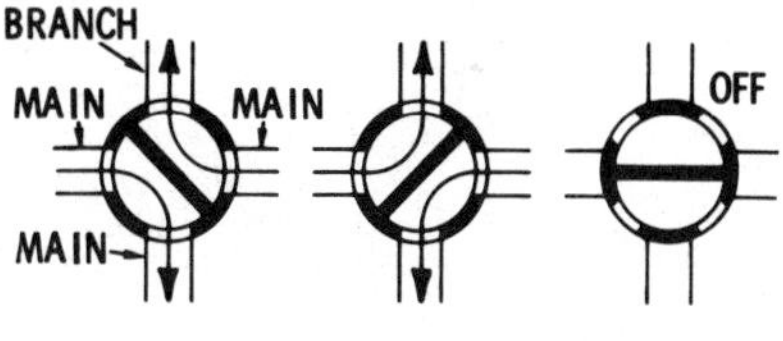

Fig. 9-37. A four-part four-way cock.

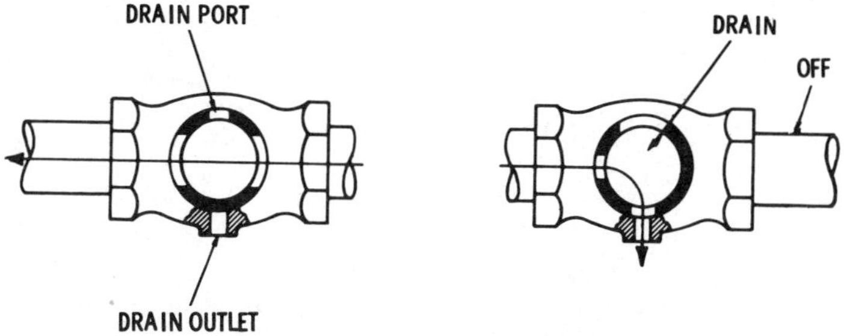

Fig. 9-38. A waste or drain cock, showing operations.

which a small or auxiliary port is provided at right angles to the two main ports, and which is connected to a drain outlet in the side of the unit. The operation of this type cock is shown in Fig. 9-38.

Waste or drain cocks should always be used to protect exposed lines in freezing weather, so that they may be conveniently shut off and drained in one operation. Corporation cocks are special forms used to shut off the water supply from a city main to a house main.

Index

A

Acrylonitrile-butadiene-styrene (ABS), 20
Adjustable wrenches, 54
Angle valves, 263
Architectural plan, 170, 171
Asphaltum, 12
Assembly, pipe, 154
  joint compound, 155
  joint make-up, 155
Awls, scratch, 33

B

Back- and side-outlet return bends, 73
Ball valves, 266
Basin, 213
Basin wrench, 54
Bathtubs
  drain hardware, 227
  fittings, 225
  roughing-in measurements, 227
Bench mark, 124
Bibb, 280
Bidet, 253
Blind flange, 78
Blow-out, 231
Blueprints, 119-135
Bonnet, 63
Bowl, 213
Branch, 63
  ell, 63
  pipe, 63
  tee, 63
Brass, 9
  fittings, 67
  pipe, 16
Bronze and corrosion-resistant flanges, 80
Building drain elevation, 132
Bull-head tee, 63
Bushings, 63, 72

C

Caps, 77
Cast-iron, 7
  fittings, 66
  flanges, 81, 84
  pipe, 15, 113
Cast-steel fittings, 67
Chain vise, 36
Chain wrench, 51
Check valves, 276
Chisels, 36
Close nipple, 63
Coated cast-iron pipe, 113
Cocks
  four-way cocks, 289
  straight-way cocks, 285
  three-way cocks, 287
  waste or drain cocks, 289
Compound wrench, 51
Compression
  faucets, 280
  joint, 110
  stops, 267
Concentric bushings, 73
Copper, 9
  pipe, 16
  tubing, 16, 162
    flexible tubing, 162
    rigid tubing, 165
Corrosion, 159
Corrosion-resistant flange, 80
Corrosive-waste pipe, 114
  glass, 115

Corrosive-waste pipe—cont.
  noncorrosive pipe, 114
  plastic, 114
Couplings, 63, 70
Cross, 63, 76
  valves, 64, 266
Crosslinked polyethylene (CAB-XL), 21
Crossover, 63
  tee, 63
Crotch, 64
Cutting soil pipe, 107

D

Double extra strong pipes, 15
Drain
  cleaner, portable electric, 56
  elevation, 132
Drainage fittings, 89
Drawing
  isometric, 177, 179
  rough-in, 170
Drop ell, 64
Drop tee, 64
Dry joint, 64

E

Eccentric bushings, 73
Elbow (ell), 64
  constants, 153
Elevations, 124-135
  using an instrument level, 132
End pattern pipe wrench, 51
Engineer's rod, 133, 134
Engineer's rule, 134
Expansion joints, 87
Extension coupling, 70
External pipe-thread dimensions, 14
Extra heavy, 64
Extra strong pipes, 15

F

Factory-made pipe heads, 160
Faucets
  compression faucets, 280
  connections, 216
  freeze-proof faucets, 283
  mixing faucets, 282
  single-lever faucets, 283
Fiberglass tub, 224
Files, 36
Fillet thermoplastic pipe welding, 27
Fittings, pipe, 61-98
  brass, 67
  bushings, 72
  cast-iron, 66
  cast-steel, 67
  definitions, 63-66
  expansion joints, 87
  forged steel, 68
  malleable iron, 66
  return bends, 76
  various, 68-98
    back- and side-outlet return bends, 75
    caps, 77
    couplings, 70
    crosses, 76
    drainage fittings, 89
    flanges, 78
    lock nuts, 68
    nipples, 68
    plugs, 77
    side-outlet elbows, 75
    soil pipe and fittings, 89
    special for wall-hung toilets, 94
    tees, 75
    unions, 71
      elbows and tees 87
    wyes, 76
Fixtures, 213-257
  bathtubs, 220
    drain hardware, 227
    fittings, 225-227
    roughing-in measurements, 225
  bidets, 253
  lavatories, 213
    kitchen sinks, 218
    roughing-in measurements, 217
    supply connections, 215
    waste connections, 216
  toilets, 230
    bowls, 230
    tanks, 235
  urinals, 241
    individual wall, 242
    pedestal, 245
    stall, 246
Flanged
  fittings and valves, 78
  swing check valve, 278
Flanges, 78
  sizes and dimensions, 79
  templates, 81-87

Flaring tools, 43
Flat threads, 146
Flat-head straight-way cock, 288
Float valves, 271
Flush valves, 275
Flux, 58
Forged-steel fittings, 68
Four-way cocks, 289
Freeze-proof faucets, 284

G

Gas torches, 59
Gate valves, 259
  non-rising stem valves, 260
  rising stem valves, 260
Glass
  pipe, 29, 115
  reinforced epoxy, 21
Globe valves, 260
Ground-joint union, 72

H

Hand
  operated threaders, 143
  stock and dies, 143
Hard steel, 8
Header, 64
Heating system rough-in, 182
Height of instrument (H.I.), 133
Horizontal life check valve, 278
Hub-and-spigot joint, 64
Hydraulic cast-iron flanges, 84
Hydrostatic joint, 64

I

Individual wall type urinals, 242
Inside-reading rule, 31
Instrument level, use of, 132
Iron
  cast, 7
  malleable, 12
Isometric drawing, 177, 179

L

Lavatories, 213
  roughing-in measurements, 217
  supply connections, 215
  waste connections, 216
Lead, 10-11
  and oakum joints, 101

Lead-cont.
  and oakum joints
    caulking the lead, 106
    pouring the lead, 103
  joint, 64
  poisoning, 10
  wool, 64
Leaky joints, 14
Levels, 33, 132
Lip union, 65
Lock nuts, 69

M

Malleable iron, 12
  fittings, 66
Malleable return bends, 74
Manifold, 65
Marking awls, 32
Material, roughing-in, 179
Measuring
  rules, 31
  tapes, 32
Mechanical plans, 170, 171
Mechanical symbols, 124
Medium
  pressure, 65
  steel, 8
Metric-standard conversion guide, 134
Mixing faucets, 282

N

Needle valves, 65, 270
Nipple, 65, 69
NO-HUB® joint, 112
Non-rising stem valves, 260

O

Oakum, 12
  and lead requirements
    for caulked joints, 104
Offsets, calculating, 149

P

Pedestal-type urinals, 246
Pipe
  and bolt taps, 49
  and tube properties, 18
  assembly, 154
  bends, 162
    bending tools, 42
    factory made bends, 160

Pipe-cont.
black, 13
brass, 16
cast-iron, 15
copper, 16
corrosion, 159
cutting, 137
cutters, 37
expansion, 159
fitting, 137-167
assembling, 154
joint compound, 155
joint make-up, 155
cutting, 137
examples, 156
tapping, 147
threading, 141
glass, 29, 115
joints, 99-117
coated cast-iron pipe, 113
compression, 110
corrosive-waste pipe, 114
cutting soil pipe, 107
lead and oakum, 100
NO-HUB®, 112
soil pipe, 99
lead, 10
plastic, 18, 168
acrylonitrite-butadiene-styrene (ABS), 20
cross linked polyethylene (CAB-XL), 21
glass reinforced epoxy, 21
polyethylene, 20
polypropylene, 20
polyvinyl chloride (PVC), 19
polyvinyl dichloride (PVDC), 20
reamer, 41
sizes, 159
supports, 158
tapping, 147
test pressures, 17
threaded joints, 13
leaky, 14
threading, 141
dimensions, external, 14
flat threads, 146
hand operated threaders, 143
power threading, 144
threaders, 44
threading machines, 47
threading operation, 146
tubes, 17
wrenches, 51

Pipe-cont.
wrenches
chain wrench, 51
compound wrench, 51
end pattern wrench, 51
straight pattern wrench, 51
vise, 138
Plan
architectural, 170, 171
mechanical, 170, 171
Plastic piping, 18, 168
acrylonitrite-butadiene-styrene (ABS), 20
bending, 23
connections, 21
crosslinked polyethylene (CAB-XL), 21
cutting, 23
expansion, 28
fillet welding, 25, 27
glass reinforced epoxy, 21
handling and storing, 21
pipe, 114, 168
polyethylene, 20
polypropylene, 20
polyvinyl chloride (PVC), 19
polyvinyl dichloride (PVDC), 20
solvent welding, 24
supporting, 29
underground installation, 24
Pliers, 50
Plugs, 77
Plumb bobs, 34
Plumbing materials, 7-30
Polyethylene, 20
Polypropylene, 20
Polyvinyl chloride (PVC), 19
Polyvinyl dichloride (PVDC), 20
Power threaders, 144
Power tools, 55
hacksaw cutting chart, 58
Pressure and temperature relief check valve, 279
Pressure regulating valves, 270

R

Radii of standard pipe bends, 161
Reducers, 65, 70
Relief valves
safety, 279
temperature and pressure, 279
Return bends, 73

Reverse-trap, 232
Right and left nipples, 72
Rising stem valves, 260
Rod threading machines, 47
Roof increaser, 65
Roughing-in, 169-212
drawing, 170
heating system, 182
isometric drawing, 177
measurements
bathtub, 227
lavatory, 217
Rules (Measuring), 31
Run, 65

S

Saws, 35
Scratch awls, 33
Service pipe, 65
Shoulder nipple, 65
Side-outlet elbows, 75
Single-lever faucets, 283
Single-wheel pipe cutter, 139
Siphon-jet, 230
Soft steel, 8
Soil pipe, 99
and fittings, 89
cutters, 40
cutting, 107
Soldering
hard, 58
soft, 58
tools, 58
torches and lead-melting devices, 59
Square bathtub, 223
Square-head straight-way cock, 288
Stall-type urinals, 246
Standard pipes, 15
Standard pressure, 65
Standard rule, 134
Steam pipe expansion, 91
Steel, 8
flanges, 82, 86
hard steel, 8
medium steel, 8
pipe, water flow, 163
soft steel, 8
Stop and waste valves, 267
Straight pattern pipe wrench, 51
Straight-way cocks, 288
Street
elbow, 65
Street-cont.
tee, 65
Supply connections
lavatory, 215
Supports, pipe, 158
Swing check valves, 278

T

T-handle tap wrenches, 48
Tables
approximate length of usable pipe thread, 156
closet elbows, dimensions, 182
crosses, dimensions, 199
reducing crosses, dimensions, 200
drainage fittings, dimensions, 210
drill sizes for standard pipe taps, 149
elbow constants, 153
expansion of steam pipes, 91
external pipe thread dimensions, 14
fillet thermoplastic pipe welding chart, 27
5⅝° and 60° elbows, dimensions, 188
45° elbows, dimensions, 186
45° Y's, dimensions, 195
reducing 45° Y's, dimensions, 196
increases and offsets, dimensions, 205
long-turn TY, dimensions, 202, 203
malleable return bends, 74
miscellaneous fittings, dimensions, 201, 204
90° elbows, dimension, 185
oakum and lead requirements for caulked joints, 104
pipe test pressures, 17
power hacksaw cutting chart, 58
properties of 1-inch tube and pipe, 18
radii and tangents of standard pipe bends, 161
60° Y's and T's, dimensions, 197
reducing 60° Y's and T's, dimensions, 198
sizes and dimensions of typical metal flanges, 79
special elbows, dimensions, 189, 190, 191
special TY's, dimensions, 206
special TY's and couplings, 207
standard malleable-iron fittings, dimensions, 157

Tables—cont.
TY's, dimensions, 193
reducing 45° TY's, 194
tap and drill sizes, 51
tapping and dimensions, 211, 212
templates,
for bronzes and corrosion-resistant flanges, 80
for 150- and 300-lb. steel flanges, 82-83
for 125-lb. cast-iron flanges, 81
for 1500- and 2500-lb. steel flanges, 86-87
for 250- and 800-lb. hydraulic cast-iron flanges, 84-85
traps, dimensions, 208, 209
22½° and 11¼° elbows, dimensions, 187
water flow in steel pipe, 163
Tangents of standard pipe bends, 162
Tapes, measuring, 32
Tap
drill sizes, 50, 51, 149
pipe and bolt, 49
wrenches, 48
Tapping, pipe, 147
Tee, 66, 75
Tee-head straight-way cock, 288
Temperature and pressure relief valve, 279
Test pressures, pipe, 17
Threaded pipe joints, 13
Threading, 141
flat threads, 146
hand operated threaders, 143
machines, 47
power threading, 144
threading operations, 146
Three-way cocks, 288, 289
Tin, 11
Toilet, 230
bowls, 230
blow-out, 231
reverse-trap, 232
siphon-jet, 230
washdown, 232
tank, 235-241
float, 239
float valves, 238
flush valves, 238
tank condensation, 239
Tools, basic, 31-60
Tubes, 17
Tubing, copper, 16, 162
cutters, 36
flexible, 162
properties, 16
reamers, 36
rigid, 166

U

Unions, 66, 71
elbows (ells), 66, 87
tee, 66, 87
Urinals, 241
individual wall-type, 242
pedestal-type, 246
stall-type, 246

V

Valves and Faucets, 259-279
angle, 263
ball, 266
check, 276
compression stops, 267
cross, 266
float, 271
flush, 275
gate, 259
globe, 260
needle, 270
pressure regulating, 270
safety relief, 279
stop and waste, 267
Vertical lift check valve, 278
Vise stand, 36

W

Wall-hung toilets, special fittings, 94
Wall-type urinals, 242
Washdown, 232
Waste connections
lavatory, 216
Waste or drain cocks, 289
Water flow or steel pipe, 163
Wyes, 66, 76

AUTOMOTIVE AIR CONDITIONING

You can easily perform most all service procedures you've been paying for in the past. This book covers the systems built by the major manufacturers, even after-market installations. Contents: introduction—refrigerant—tools—air conditioning circuit—general service procedures—electrical systems—the cooling systems—system diagnosis—electrical diagnosis—troubleshooting. 232 pages; $5\frac{1}{2} \times 8\frac{1}{4}$; softcover. **Price: $7.95**

DIESEL ENGINE MANUAL, 4th Edition

A practical guide covering the theory, operation and maintenance of modern diesel engines. Explains diesel principles—valves—timing—fuel pumps—pistons and rings—cylinders—lubrication—cooling system—fuel oil and more. 480 pages; $5\frac{1}{2} \times 8\frac{1}{4}$; hardbound. **Price: $12.95**

GAS ENGINE MANUAL, 2nd Edition

A completely practical book covering the construction, operation, and repair of all types of modern gas engines. 400 pages; $5\frac{1}{2} \times 8\frac{1}{4}$; hardbound. **Price: $9.95**

SMALL GASOLINE ENGINES

A new manual providing practical and theoretical information for those who want to maintain and overhaul two- and four-cycle engines such as lawn mowers, edgers, snowblowers, outboard motors, electrical generators, and other equipment using engines up to 10 horsepower. 624 pp; $5\frac{1}{2} \times 8\frac{1}{4}$; hardbound. **Price: $15.95**

TRUCK GUIDE—3 Vols.

Three all-new volumes provide a primary source of practical information on truck operation and maintenance. Covers everything from basic principles (truck classification, construction components, and capabilities) to troubleshooting and repair. 1584 pages; $5\frac{1}{2} \times 8\frac{1}{4}$; hardbound.
Price: $41.85

Volume 1
ENGINES: **$14.95**
Volume 2
ENGINE AUXILIARY SYSTEMS: **$14.95**
Volume 3
TRANSMISSIONS, STEERING AND BRAKES: **$14.95**

BUILDING AND MAINTENANCE

ANSWERS ON BLUEPRINT READING, 3rd Edition

Covers all types of blueprint reading for mechanics and builders. This book reveals the secret language of blueprints, step by step in easy stages. 312 pages; $5\frac{1}{2} \times 8\frac{1}{4}$; hardbound.
Price: $9.95

BUILDING MAINTENANCE, 2nd Edition

Covers all the practical aspects of building maintenance. Painting and decorating; plumbing and pipe fitting; carpentry; heating maintenance; custodial practices and more. (A book for building owners, managers, and maintenance personnel.) 384 pages; $5\frac{1}{2} \times 8\frac{1}{4}$; hardbound.
Price: $9.95

COMPLETE BUILDING CONSTRUCTION

At last—a one volume instruction manual to show you how to construct a frame or brick building from the footings to the ridge. Build your own garage, tool shed, other outbuildings—even your own house or place of business. Building construction tells you how to lay out the building and excavation lines on the lot; how to make concrete forms and pour the footings and foundation; how to make concrete slabs, walks, and driveways; how to lay concrete block, brick and tile; how to build your own fireplace and chimney. It's one of the newest Audel books, clearly written by experts in each field and ready to help you every step of the way. 800 pages; $5\frac{1}{2} \times 8\frac{1}{4}$; hardbound. **Price: $19.95**

Use the order coupon on the back of this book.

All prices are subject to change without notice.

Volume 2
Bricklaying, Plastering Rock Masonry, Clay Tile. 384 pages; 5½ × 8¼; hardbound.
Price: $12.95

PAINTING AND DECORATING
This all-inclusive guide to the principles and practice of coating and finishing interior and exterior surfaces is a fundamental sourcebook for the working painter and decorator and an invaluable guide for the serious amateur or building owner. Provides detailed descriptions of materials, pigmenting and mixing procedures, equipment, surface preparation, restoration, repair, and antiquing of all kinds of surfaces. 608 pages; 5½ × 8¼; hardbound. **Price: $18.95**

PLUMBERS AND PIPE FITTERS LIBRARY, 3rd Edition—3 Vols.
A practical, illustrated trade assistant and reference for master plumbers, journeymen and apprentice pipe fitters, gas fitters and helpers, builders, contractors, and engineers. Explains in simple language, illustrations, diagrams, charts, graphs, and pictures the principles of modern plumbing and pipe-fitting practices. **Price: $32.85**

Volume 1
Materials, tools, roughing-in. 320 pages; 5½ × 8¼; hardbound. **Price: $11.95**
Volume 2
Welding, heating, air-conditioning. 384 pages; 5½ × 8¼; hardbound. **Price: $11.95**
Volume 3
Water supply, drainage, calculations. 272 pages; 5½ × 8¼; hardbound. **Price: $11.95**

THE PLUMBERS HANDBOOK, 7th Edition
A pocket manual providing reference material for plumbers and/or pipe fitters. General information sections contain data on cast-iron fittings, copper drainage fittings, plastic pipe, and repair of fixtures. 330 pages; 4 × 6 softcover. **Price: $9.95**

QUESTIONS AND ANSWERS FOR PLUMBERS EXAMINATIONS, 2nd Edition
Answers plumbers' questions about types of fixtures to use, size of pipe to install, design of systems, size and location of septic tank systems, and procedures used in installing material. 256 pages; 5½ × 8¼; softcover. **Price: $8.95**

TREE CARE MANUAL
The conscientious gardener's guide to healthy, beautiful trees. Covers planting, grafting, fertilizing, pruning, and spraying. Tells how to cope with insects, plant diseases, and environmental damage. 224 pages; 8½ × 11; softcover. **Price: $8.95**

UPHOLSTERING
Upholstering is explained for the average householder and apprentice upholsterer. From repairing and regluing of the bare frame, to the final sewing or tacking, for antiques and most modern pieces, this book covers it all. 400 pages; 5½ × 8¼; hardbound. **Price: $12.95**

WOOD FURNITURE: Finishing, Refinishing, Repair
Presents the fundamentals of furniture repair for both veneer and solid wood. Gives complete instructions on refinishing procedures, which includes stripping the old finish, sanding, selecting the finish and using wood fillers. 352 pages; 5½ × 8¼; hardbound. **Price: $9.95**

ELECTRICITY/ELECTRONICS
ELECTRICAL LIBRARY
If you are a student of electricity or a practicing electrician, here is a very important and helpful library you should consider owning. You can learn the basics of electricity, study electric motors and wiring diagrams, learn how to interpret the NEC, and prepare for the electrician's examination by using these books.

Use the order coupon on the back of this book.
All prices are subject to change without notice.

Electric Motors, 4th Edition. 528 pages; $5\frac{1}{2} \times 8\frac{1}{4}$; hardbound. **Price: $12.95**

Guide to the 1984 National Electrical Code. 672 pages; $5\frac{1}{2} \times 8\frac{1}{4}$; hardbound. **Price: $18.95**

House Wiring, 6th Edition. 256 pages; $5\frac{1}{2} \times 8\frac{1}{4}$; hardbound. **Price: $12.95**

Practical Electricity, 4th Edition. 496 pages; $5\frac{1}{2} \times 8\frac{1}{4}$; hardbound. **Price: $13.95**

Questions and Answers for Electricians Examinations, 8th Edition. 288 pages; $5\frac{1}{2} \times 8\frac{1}{4}$; hardbound. **Price: $12.95**

ELECTRICAL COURSE FOR APPRENTICES AND JOURNEYMEN, 2nd Edition

A study course for apprentice or journeymen electricians. Covers electrical theory and its applications. 448 pages; $5\frac{1}{2} \times 8\frac{1}{4}$; hardbound. **Price: $13.95**

FRACTIONAL HORSEPOWER ELECTRIC MOTORS

This new book provides guidance in the selection, installation, operation, maintenance, repair, and replacement of the small-to-moderate size electric motors that power home appliances and over 90 percent of industrial equipment. Provides clear explanations and illustrations of both theory and practice. 352 pages; $5\frac{1}{2} \times 8\frac{1}{4}$; hardbound. **Price: $15.95**

TELEVISION SERVICE MANUAL, 5th Edition

Provides the practical information necessary for accurate diagnosis and repair of both black-and-white and color television receivers. 512 pages; $5\frac{1}{2} \times 8\frac{1}{4}$; hardbound. **Price: $15.95**

ENGINEERS/MECHANICS/MACHINISTS

MACHINISTS LIBRARY, 4th Edition

Covers the modern machine-shop practice. Tells how to set up and operate lathes, screw and milling machines, shapers, drill presses and all other machine tools. A complete reference library. **Price: $35.85**

Volume 1
Basic Machine Shop. 352 pages; $5\frac{1}{2} \times 8\frac{1}{4}$; hardbound. **Price: $12.95**

Volume 2
Machine Shop. 480 pages; $5\frac{1}{2} \times 8\frac{1}{4}$; hardbound. **Price: $12.95**

Volume 3
Toolmakers Handy Book. 400 pages; $5\frac{1}{2} \times 8\frac{1}{4}$; hardbound. **Price: $12.95**

MECHANICAL TRADES POCKET MANUAL, 2nd Edition

Provides practical reference material for mechanical tradesmen. This handbook covers methods, tools equipment, procedures, and much more. 256 pages; 4×6; softcover. **Price: $10.95**

MILLWRIGHTS AND MECHANICS GUIDE, 3rd Edition

Practical information on plant installation, operation, and maintenance for millwrights, mechanics, maintenance men, erectors, riggers, foremen, inspectors, and superintendents. 960 pages; $5\frac{1}{2} \times 8\frac{1}{4}$; hardbound. **Price: $19.95**

POWER PLANT ENGINEERS GUIDE, 3rd Edition

The complete steam or diesel power-plant engineer's library. 816 pages; $5\frac{1}{2} \times 8\frac{1}{4}$; hardbound. **Price: $16.95**

WELDERS GUIDE, 3rd Edition

This new edition is a practical and concise manual on the theory, practical operation and maintenance of all welding machines. Fully covers both electric and oxy-gas welding. 928 pages; $5\frac{1}{2} \times 8\frac{1}{4}$; hardbound. **Price: $19.95**

Use the order coupon on the back of this book.

All prices are subject to change without notice.

WELDER/FITTERS GUIDE

Provides basic training and instruction for those wishing to become welder/fitters. Step-by-step learning sequences are presented from learning about basic tools and aids used in weldment assembly, through simple work practices, to actual fabrication of weldments. 160 pages; $8^1/_2 \times 11$; softcover. **Price: $7.95**

FLUID POWER

PNEUMATICS AND HYDRAULICS, 4th Edition

Fully discusses installation, operation and maintenance of both HYDRAULIC AND PNEUMATIC (air) devices. 496 pages; $5^1/_2 \times 8^1/_4$; hardbound. **Price: $15.95**

PUMPS, 4th Edition

A detailed book on all types of pumps from the old-fashioned kitchen variety to the most modern types. Covers construction, application, installation, and troubleshooting. 480 pages; $5^1/_2 \times 8^1/_4$; hardbound. **Price: $14.95**

HYDRAULICS FOR OFF-THE-ROAD EQUIPMENT

Everything you need to know from basic hydraulics to troubleshooting hydraulic systems on off-the-road equipment. Heavy-equipment operators, farmers, fork-lift owners and operators, mechanics—all need this practical, fully illustrated manual. 272 pages; $5^1/_2 \times 8^1/_4$; hardbound. **Price: $8.95**

HOBBY

COMPLETE COURSE IN STAINED GLASS

Written by an outstanding artist in the field of stained glass, this book is dedicated to all who love the beauty of the art. Ten complete lessons describe the required materials, how to obtain them, and explicit directions for making several stained glass projects. 80 pages; $8^1/_2 \times 11$; softbound. **Price: $6.95**